# TADAO ANDO
## Shanghai Poly Grand Theatre
安藤忠雄 | 上海保利大剧院

Architecture and Urbanism
July 2015 Special Issue
建筑与都市
2015 年 7 月专辑

A+U Publishing Co., Ltd
A+U 出版社

Publisher/Chief Editor: Nobuyuki Yoshida

Executive Chief Editor: MA Weidong

Planning Editor: Kazuya Okano

Editor: WAN Ying

Assistant Editors: WU Ruixiang
　　　　　　　　 CHEN Xudan

发行人 / 主编：吉田信之

执行主编：马卫东

策划编辑：冈野一也

编　　辑：完　颖

助理编辑：吴瑞香　陈旭丹

Tongji University Press
同济大学出版社

Producer: ZHI Wenjun

Executive Editor: HU Yi

Assistant Editor: LV Wei

Executive Proofreader: XU Chunlian

出 品 人：支文军

责任编辑：胡　毅

助理编辑：吕　炜

责任校对：徐春莲

Cover: Local residents participating in various activities at the semi-outdoor cylindrical space
封面：市民在剧院半室外圆筒空间内活动的场景

| 28 | **Chapter 1: Design Process** | | |
|---|---|---|---|
| | 30 | Shanghai | |
| | 31 | Jiading | |
| | 32 | Jiading New City | |
| | 34 | Site | |
| | 38 | Site Analysis: Theatre | |
| | 40 | Program | |
| | 48 | Preliminary Study Process | |
| | 52 | Plans | 54 Plan A |
| | | | 55 Plan B |
| | 62 | Design Concept | 56 Plan C |
| | 64 | Formal Concept | |
| | 68 | Spatial Composition | |
| | 80 | Design Concept | 82 Entrance Hall |
| | 136 | Drawings | 88 Auditorium |
| | | | 90 Visual Line Analysis |
| | | | 92 Traffic Streamline, Fire Safety, and Greenery Analysis |
| | | | 94 Landscape Analysis |

| 152 | **Chapter 2: Features** |
|---|---|
| | 154　Fair-faced Concrete |
| | 158　Curtain Wall |
| | 164　Cylinder Nodes |
| | 170　Interior Decoration of the Auditorium |
| | 174　Lighting |
| | 178　Theatre Acoustics |

| 208 | **Chapter 3: Technological Realization** |
|---|---|
| | 214　Plan Drawings |
| | 216　Fair-faced Concrete |
| | 220　Interior Decoration |
| | 222　Stage Mechanics |

| Essay | 24 | The Challenges of Creating a Cathedral to Culture/Tadao Ando |
|---|---|---|
| | 42 | Lighting up the City/Interview with SUN Jiwei |
| | 58 | Cultural Kaleidoscope/Interview with Tadao Ando |
| | 96 | Dynamic Tubes: Ando's Vision Realized in a New Urban Community/Kazukiyo Matsuba |
| | 182 | The Theatre and Me/ZHAO Guo-ang |
| | 210 | Pearls Strewn in Concrete and Soil/SUN Jian |

| Photo | 6 | Photo of Exterior |
|---|---|---|
| | 104 | Photo of Interior |
| | 184 | Photo of Under Construction |

| | 224 | Data |
|---|---|---|
| | 226 | Chronology |

| 28 | 第1章：方案演绎 |
|---|---|

- 30 上海
- 31 嘉定
- 32 嘉定新城
- 34 基地
- 38 剧院分析
- 40 项目内容
- 48 初期探讨
- 52 方案评述
  - 54 方案 A
  - 55 方案 B
  - 56 方案 C
- 62 设计理念
- 64 形体概念
- 68 空间构成
- 80 设计方案
  - 82 入口大厅
  - 88 观众厅
  - 90 视线分析
  - 92 交通流线、消防、绿化分析
  - 94 景观分析
- 136 图纸

| 152 | 第2章：特点构成 |
|---|---|

- 154 清水混凝土
- 158 幕墙
- 164 圆筒节点
- 170 观众厅室内装饰
- 174 照明
- 178 剧场声学

| 208 | 第3章：技术实现 |
|---|---|

- 214 方案的图纸实现
- 216 清水混凝土实现
- 220 室内装饰实现
- 222 舞台机械实现

文章
- 24 文化殿堂的挑战 / 安藤忠雄
- 42 建筑点亮城市 / 采访孙继伟
- 58 文化万花筒 / 采访安藤忠雄
- 96 跃动的"圆筒"——凝筑于新城中的安藤理念 / 松叶一清
- 182 剧院·缘 / 赵国昂
- 210 散落在营造中的珠玑 / 孙健

照片
- 6 大剧院外观照片
- 104 大剧院室内照片
- 184 大剧院施工照片

- 224 数据
- 226 年表

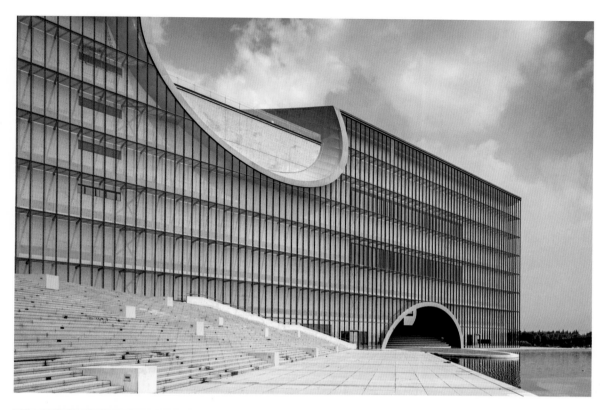

PP6-7: The elevation view of Shanghai Poly Grand Theatre
PP8-9: The view of Shanghai Poly Grand Theatre from the Yuanxiang Lake
PP10-11: Multiple cylinders intersect the cuboid theatre, endowing the space with greater nuance

6-7 页：上海保利大剧院侧立面
8-9 页：从远香湖湖面远眺上海保利大剧院
10-11 页：数个圆筒穿插于立方体剧院之中，使空间层次更为丰富

PP12-13: The cylindrical spaces create ovoid openings by extending into and crisscrossing with the exterior rectangular façade
PP14-15: The lights light up brilliantly at Shanghai Poly Grand Theatre
PP16-17: The Grand Theatre's public spaces are another stage for visitors
PP18-19: Local residents congregate for night-time activities at the Grand Theatre's plaza, giving the impression of performers dancing on stage
PP20-21: Variations in the ovoid openings add nuance to the space
PP22-23: The amphitheatre is like a moon, whose graceful ascent is reflected in the water's gentle ripples

12-13页：圆筒空间的延伸，与长方体外部的幕墙交错形成了椭圆形开口
14-15页：华灯初上的上海保利大剧院
16-17页：大剧院公共空间部分是留给观众的另一个舞台
18-19页：夜晚，居民在剧院广场的活动，使得他们好像表演者般在舞台上婆娑起舞
20-21页：圆筒状的开口变化，创造出不同的空间感受
22-23页：半室外舞台倒映在水中，犹如海上升起一轮明月

# The Challenges of Creating a Cathedral to Culture
文化殿堂的挑战

Tadao Ando　安藤忠雄

**Designing in China**

It was in 2005 that I began working in China in earnest. Right around that time I had an architecture exhibition in Shanghai, which, being the dynamic heart of a new China, is a frenetic and constantly changing city. In presenting my ideas on architecture in this bustling city at the height of its growth, I elected to focus on the theme of the "environment." Part of my message was that "growth does not simply mean 'out with the old, in with the new'… different places have different contexts, and each city has its own unique ideal to strive toward." In a way this reflected some of my anxieties about commencing work as an architect in Shanghai. I was concerned that in the process of designing structures for a city in an impetuous rush to grow, where everything is swept along by the tide of the urban economy and where there is no time to stop and think, I would be unable to maintain my insistence on creating unique buildings tailored to specific sites. A decade later, as of 2014, I have ten projects in China, primarily in Beijing and Shanghai, each of which has its own daunting challenges but is moving forward nonetheless. Among them is the Shanghai Poly Grand Theatre, for which construction began in 2009 and is now about to be completed. The project is an opera house complex that is intended to serve as the cultural nexus of a newly developed community on the outskirts of Shanghai. Originally, the construction period allotted for this immense project was three years, which is quite typical of China's features. I was a bit incredulous at the start, and sure enough it ended up taking five years, but it was successfully completed almost exactly as I had environed it—a spectacular feat of engineering. To me, however, what was most meaningful and rewarding was the process of getting it built.

**Designing the Opera House**

The Shanghai Poly Grand Theatre was planned as the cultural centre of Jiading New City (approximately 220km$^2$) of Jiading District, a suburb of Shanghai where urban development is proceeding at a breakneck pace. The allocated site was part of a lakeside park in the centre of the area under development, and the facilities to be built were divided into a commercial zone including offices and a hotel, as well as a cultural zone featuring the opera house as its centrepiece. Naturally, designing the opera house would be the main challenge of this project.
Opera is a multimedia art form that encompasses music, theatre, and the visual arts. For Western Europeans, opera has held a central position in the fine arts since the 17th century; cities throughout the region have staked their prestige on building grand opera houses for the presentation of this art form to the public. Preserving these buildings in all their glory for future generations as a sacred mission for these cities.
Take, for example, the Teatro La Fenice [1] in Venice, Italy, which has been burned to the ground twice, in 1836 and in 1996. However, true to its name (which means "the Phoenix"), both times it has arisen from the ashes, remaining one of the world's most renowned and beloved venues for opera. In 2003, when I was in northern Italy for work reasons, I was lucky enough to attend the first performance at La Fenice after its second reconstruction. Over a period of eight years, under the determined leadership of Venice's then-mayor Paolo Costa, it was rebuilt "how it was, where it was," faithfully replicating the Neo-Baroque interior of the original while also incorporating the latest state-of-the-art stage equipment. I was awed by the power and grandeur of the lovingly restored space.
Right at the same time, the Teatro alla Scala [2] in Milan was in the midst of a major renovation. The architect in charge was my friend Mario Botta. When I dropped by to visit him and to see how things were going, he brought me to the site to view the work in progress. I listened to Botta talking with tremendous gusto about what an honor it was to direct this renovation project, and what sort of ideas he was working with as he strove to meet people's soaring expectations, I got a palpable sense of just how much an opera house means to a Western European architect.
As for me, I had no shortage of experience designing museums and had my own ideas about museum architecture, but I had never been involved with an opera house project before. As a public cathedral to culture, what sort of space was an opera house supposed to be? I turned the question over and over in my mind, and finally I got a chance to come up with my own answer to it, in the form of the Shanghai Poly Grand Theatre.
Naturally, the greatest challenge in planning an opera house is designing the aspects critical to the site's functionality, particularly the acoustics. An exacting technical approach is required to achieve the reverberation frequencies appropriate to the pieces being performed, while allowing for suitable positioning of audience seating. Another thing that I thought was important, and which contributes to the drama of the overall experience, is the sequence of spaces that people experience on their way to their seats.
Take the hall of the grand staircase at the Palais Garnier in Paris, for example. The interior, with a high vaulted ceiling and a four-storey wraparound gallery, has all the dynamism and grandeur of the Baroque era, and could well be called an opera stage in its own right. Spirits soar and everyday worries are forgotten as the audience enters the auditorium and approaches the stage, a sensation that is heightened by Marc Chagall's ceiling mural, which overlooks the entire space. Soon, the audience is immersed in the mythic realm of the narratives being presented on stage.
My main theme at the Shanghai Poly Grand Theatre was to design my kind of structure using the techniques of contemporary architecture, while delivering the riches and majesty that an opera house ought to have. The solution I arrived at can be observed in the structure's spatial composition: a series of very large cylindrical tubes that slice through the Theatre's rectangular space.

## 在中国的工作

说起在中国正式开展工作,那要追溯到 2005 年,那时我在上海举办了一场建筑展览。上海,作为新时代中国的中心城市,每天都以充沛的精力变化发展着。面向上海这座发展得如火如荼的城市,我选择了以"环境"为主题展示我的建筑理念。

在那次展览中,我传递了一个信息,即单纯地以新事物覆盖旧事物的方式,根本不能称之为成长……每一个具有不同特点的地方,都应有其固有的所追求的城市形态。话虽如此,这也正反映了我的担忧,当时我正准备开始在中国做项目,我一直怀疑,城市经济发展的势头将会吞没一切,在所有事情都一刻不停地向前推动的大环境下,我是否仍能坚守自身的想法,坚持以"某地独一无二、度身而造的建筑"为不变的主题来做设计。转眼已过去十年,直至 2014 年我已经在北京、上海等多个城市承接了十多个项目,这些项目无一不是通过艰苦卓绝的努力而向前推进着。这其中开始于 2009 年的"上海保利大剧院"项目也终于迎来了竣工的好消息。

这是一座包括剧院在内的文化综合设施,她的使命是要成为上海近郊嘉定新城的文化中心。虽然项目规模如此巨大,但当初给我们的建设时间却只有三年多,包含了从开始设计到项目竣工整个过程,这倒颇符合中国式的做法。我是怀着半信半疑的心情开始了设计,虽然最终花费了五年时间,但建成后的项目基本还原了我最初的设计构想,应该说是一件非常了不起的事情。然而,之于我,更重大的意义莫过于整个项目的实施过程给我带来的充实感。

## 以剧院为主题

嘉定区位于上海郊外,其城市建设正在飞速进行中,其中规划建设 220km² 为嘉定新城,而"上海保利大剧院"将建设成为嘉定新城的文化中心。项目基地位于开发区的中心地区,其占据了湖滨公园一角。基地包括商业地块及文化地块,其中商业地块包含酒店及办公楼,文化地块则以大剧院为中心,建成的基地区域将会是商业与文化两种功能兼具的综合文化设施。其中,最大的课题无疑是大剧院的设计。

所谓歌剧,是音乐、戏剧、舞台美术等经洗练后融为一体的综合性艺术。在西方世界里,剧院从近代以来便一直是艺术活动的中心。正因为如此,剧院作为歌剧表演的场所,每个城市都不遗余力地致力于它的建设,并以守护、传承这个建筑为无上使命。

就拿意大利威尼斯的凤凰剧院[1]来说,1836 年和 1996 年曾两次遭受

火灾,几乎全部焚烧殆尽,可它却如同它的名字一样,每次都被重建,时至今日仍然是歌剧圣地之一,受到全世界人民的爱护。2003 年,我因公出差来到北意大利,有幸造访了第二次被重建后的凤凰剧院,重建工作在当时威尼斯市长保罗·科斯塔的强力支持下,采用了最新的舞台设备,同时打出"在原址上按原样重建"的标语,内装修复工作也是持着无论如何都要恢复原始巴洛克风格的决心,历时 8 年重建完成。重生后的凤凰剧院还是那么金碧辉煌,给人强烈的力量感。

正好那时位于意大利米兰的斯卡拉歌剧院[2]也在进行全面的翻新工程,担当这项工程的建筑师马利奥·博塔是我的朋友,在我专程去拜访他时,他亲自带我参观了施工现场。他一直认为,能被委以翻新剧院的重任是自己的无上光荣,博塔热血沸腾地为我讲解了他为不辜负这种期待而绞尽脑汁尝试了何种设计方法,当时我看着他的侧脸,再次感受到了剧院建设之于西欧建筑师们而言的重要意义。

我虽然在美术馆、博物馆设计方面有不少经验,也有了一些切身体会,却一直未接触过剧院设计,其实我也一直在问自己,剧院要如何设计才能称得上是文化殿堂的空间。"上海保利大剧院"的设计让这个一直盘旋在我心中的疑问,终于得到了展示自己答案的机会。

说到剧院设计,最重要的课题无疑是要满足以声学为中心的极高的功能要求,根据预想的演出使用要求,对混响时间及观众厅座位配置等做出

---

1  Teatro La Fenice, also known as the Opera House of the Phoenix, is the most notable of its kind in Venice, and one of the most renowned in Europe. Over 200 years of age, La Fenice has been burned down and rebuilt three times. Begun in 1790, La Fenice caught fire before it was even finished. Fortunately, it was resorted afterwards and opened to the public in April 1792. In 1836, La Fenice was completely destroyed by fire and was subsequently rebuilt in 1854 and 1938. At that time, it could hold 1,500 spectators and contained 96 boxes. During its renovation in 1996, La Fenice caught fire again. It was rebuilt in 2001 after the design of the late architect Aldo Rossi and opened to the public in 2003.
2  Located in Milan, Italy, the Teatro alla Scala was built on the site of the Church of Santa Maria alla Scala in August 1778, hence its name. The magnificent and dignified La Scala is a three-storey opera house designed by the neoclassical architect Giuseppe Piermarini.

1  威尼斯凤凰歌剧院,又称不死鸟歌剧院,是威尼斯最主要的歌剧院,也是欧洲最著名的歌剧院之一,迄今已有两百多年的历史。在剧院历史上它曾三度失火,又三度重建。其中 1790 年开始兴建,但在建造工程尚未完成之时,就遭遇了火灾。后经重建,于 1792 年 4 月落成开幕。1836 年歌剧院因失火被彻底烧毁,后在 1854 年和 1938 年两次进行重建,能容纳 1,500 名观众,有 96 个包厢。1996 年,正在修复中的凤凰歌剧院再次失火,后于 2001 年进行重建,在已故意大利建筑师阿尔多·罗西的设计方案基础上,2003 年建成开幕。
2  斯卡拉歌剧院,位于意大利米兰,落成于 1778 年 8 月,由于是在圣玛利亚斯卡拉教堂的原址上修建的,而得名为斯卡拉歌剧院。斯卡拉歌剧院为三层建筑,规模宏伟,凝重大方,由新古典主义建筑师朱色佩·皮尔马利尼设计。

## Spatial Concept

The basic structural framework for the Shanghai Poly Grand Theatre is a cuboid with a 100m by 100m square base and a height of 34m. The 1,600-seat main auditorium is positioned on an axis diagonal to the square base. The axis begins at the northeast corner where the approach to the auditorium begins.

The other necessary functions of the site were to be positioned within the cuboid outside the main auditorium, and my idea for their layout revolved around a freely structured series of cylindrical ovoids which were 18m in diameter.

Cylinders in and of themselves are simple and clearly axial geometric shapes, but having them connect and intersect with one another would give rise to a complex and multifaceted sequence of spaces with surprises around every corner. It would be a public space that fosters a vibrant cultural life, another "stage" in its own right. Air and light would be introduced into each area in accordance to its function, ensuring that the "stage" is always shown at its finest. When presenting my concept, I used the phrase "cultural kaleidoscope" to convey this image of the visitors' vibrant activity inside these cylindrical tubes. The tubes, extending this way and that, would slice through the exterior curtain wall surrounding the cuboid, creating elliptical apertures when they hit the wall at oblique angles. From the outside, these boldly curved forms would hint at the intense drama lurking behind the building's tranquil façade.

## The Difficult Process of Execution

After working out the basic plan, I made a presentation in the construction site. I was taking a bit of a risk in proposing this idea, which seemed a bit introverted and contemplative, in a country that leans toward bright, orderly, and symmetrical structures reflecting its vastness and might, but the clients agreed to it without hesitation, perhaps sympathizing with my strong desire to build an opera house that was unlike any other and (I hoped) would help put Jiading District on the map. I believe the clients were also hoping for a future-oriented opera house worthy of being called a cathedral of culture in 21st-century China. After that, dialogue with the local construction team began, kicking off a hectic period during which work the logistics of the final design were being hashed out while work on the foundation was already under way. The greatest practical hurdle was the construction of the tubes at the core at my proposal. At first we considered using unfinished concrete to create powerful, continuously flowing spaces, but taking the scale into account, it was decided that this would be structurally impractical. After that, we considered using a series of rings of laminated wood to make ribbed, curved interiors for the tubes, but this idea was eventually discarded for fire safety concerns, and we settled on specially ordered aluminum ribs instead.

During the construction process, the problems we encountered were all in the tubes. Since it was based on a clear and simple architectural diagram—a series of interconnecting tubes cutting through a cuboid volume—it looked easy to build at first glance, but once work got underway this turned out not to be the case. The intersection points formed complex curves, and construction required a high degree of precision because the geometry of these curves would be fully exposed to view. The local construction team struggled mightily with the task, which would have been a challenge even for a Japanese architectural firm with a global reputation for effective oversight of elaborate construction projects. Nobody involved with the project was confident that it would actually get built as planned.

However, the local construction team, headed by a project manager on the clients' side, unflinchingly tackled this arduous task with terrific enthusiasm and scrupulous attention to detail. Every one of the ribs, which differed in length and angle depending on their location within the structure, was mapped with CAD, produced in a factory with great precision and delivered to the site. Industrial products made in factories can have flaws though, and that is part of the difficulty of creating a one-of-a-kind piece of architecture. On the Shanghai Poly Grand Theatre site, as well, there were many problems during the initial stages, with parts having the wrong dimensions and things not fitting together right, especially at the crucial joints between the tubes. Undaunted, each time the team was faced with an incorrectly manufactured part, they would make another mockup, examine the structure at its actual dimensions and make improvements, repeating this process over and over and moving forward step by step. Thanks to their tenacity and can-do spirit, the end result was a building that exceeded everyone's expectations, even those of its designer.

## The Construction Process

During the Shanghai Poly Grand Theatre project, I was reminded with renewed force that the essence of architecture is not at the drawing board but at the site, where the construction actually takes place. In the real world, bound by a rigid schedule and budget constraints, choosing a process like this one, which involved grappling with and resolving problems one by one, might be called irrational. Indeed, in the world of architectural design, the dominant trend seems to be toward cool sophistication, abstraction, and avoidance of get-your-hands-dirty work that this project entailed.

However, as the environment continues to change, it is always human beings who set out to design architecture, and human beings who build structures and occupy them. To create architecture for human beings, with all of our contradictions and imperfections, it is sometimes necessary to take persistence to the extreme, and to propose ideas that come from the impractically delicate, warm and breathing human spirit and body. There is a tension unique to spaces achieved through a difficult and protracted process of trial and error. Yet it is because of this time-consuming and laborious process that the building is imbued with real joy upon its completion.

In working on projects in various Asian countries from a home base in Japan, architects are bound to be perplexed by differences in rules, systems of management, and other practicalities of architecture, and to dwell excessively on the shortcomings. This has been true for me as well. However, the construction of the Shanghai Poly Grand Theatre would not have been possible without the brash, adventurous spirit of a country still on the path of development. I feel extremely fortunate and deeply grateful to have been engaged, as an architect, with this project, which showed me so many new possibilities that point the way to the architecture of tomorrow.

(**Tadao Ando**  Architect/Shanghai Poly Grand Theatre Senior Designer)

相应设定，这就需要极高的技术支持。除此之外，还有一点也是我认为至关重要的，那就是来到剧院的人们所能感受到的空间的演绎，也就是所谓的剧场性。

比如巴黎歌剧院壮观的大楼梯，四层楼高的画廊环绕于整个挑空部分，加之巴洛克建筑所独有的生机勃勃的空间，真可谓是剧院的另一个舞台。当你踏入这个被夏加尔的画作俯看的空间，再慢慢走入剧院内部，会不由地情绪高昂起来，并渐渐发现自己正被带入一个奇幻的童话世界。

这种剧院应有的空间的丰富性要如何通过现代建筑的手法去表现？又如何将其作为自己独创的建筑空间去表现？这即是我所设计"上海保利大剧院"这座建筑的主题。并且，我找到了属于自己的空间组成的答案——将剧院内嵌于长方体中，并通过一系列立体变化的圆筒空间将长方体切分，从而构成建筑。

## 空间设计

现已竣工的"上海保利大剧院"，其建筑框架是一个与远香湖相连的长方体。长方体底面是一个边长100m的正方形，高34m。剧院入口设置在长方体的东北端，从这个端点拉出的对角线作为剧院的主轴线，将1,600座的主剧场设置于主轴线上。

主剧场位置确定后，长方体中剩下的空间即可用以设置其他各项剧院必备功能空间；此时，我所采用的空间构成手法是用直径18m的圆筒去创造自由自在的挑空空间。

虽然圆筒形本身只是一种轴线清晰的简单的几何体，但通过圆筒的立体交错并连接这些空间后，即创造出复杂多样的空间表情，同时也孕育出意想不到的空间变化。由此形成的公共空间，可激发人们积极地进行文化活动，从而成为另一个舞台。此外，根据功能的不同，每个空间形成的风与光的演绎也构成了一个舞台，在这个丰富的舞台空间内我想要表现的是人们活动的景象，因此我选择了"文化万花筒"这个词来形容这种熠熠生辉的场景。

圆筒在长方体内部向各个方向延伸，到外围时被覆盖于长方体之外的幕墙截断，这些不同角度的截面便形成了椭圆形的开口，这些大胆的曲线外框，正是暗示了在安静的立面下潜藏着汹涌的内部空间变化。

## 艰难的实现过程

在方案设计阶段结束后我曾去现场做了一次汇报。中国一直以来都推崇大国建筑，喜欢强烈的轴线性和对称性，在这样的建筑文化土壤中，我却大胆地进行了尝试和挑战。从某种意义上来说本次方案是一个内向型的建筑方案，但是业主却毫不犹豫地采纳了我的设计，因为我们在同一点上有了共鸣——希望做出一个值得全世界赞扬，同时又是该地区独一无二的剧院。我想业主当时肯定也是着眼于未来，抱着建造出一座堪称21世纪中国文化殿堂的剧院的决心吧！

紧接着，在与当地设计团队进行交流后便开始了施工图设计与桩基工程同时进行的匆忙的日子。

在方案的实际深化过程中，遇到的最大问题，也是本次设计的核心问题，便是如何实现圆筒空间。一开始我们计划采用清水混凝土做出既有力又连续的空间方案，但考虑到结构上无法实现以及与建筑体量感之间的关系后便放弃了。之后，我们开始研究采用弧形松木集成材连续形成的格栅曲面组成圆筒的方案，但由于防火防灾上的要求，圆筒方案最终采用定做的铝方管格栅材料。

当然，进入施工阶段后，问题的焦点仍集中在圆筒上。虽然从构图上看仅仅只是矩形体块与圆筒组合而成的简单构成，然而真正开始建造后却并非如此。因为建筑要表现出那些圆筒与圆筒相交处自然描绘出来的复杂曲线以及最真实的几何学效果，需要很高的施工精度，为此现场的施工团队付出了巨大的艰辛与汗水。他们通过严谨的现场管理克服了重重困难，

哪怕是对于备受世界称道的日本施工单位来说也并非易事，毕竟每一个参与者都无法确定，这个项目最终是否能像设想的效果那样建造出来。

但是，在业主方项目经理的带领下，现场施工团队毫无畏惧，胆大心细，克服了重重困难。那些格栅材料所处不同部位就有不同的长度及角度，这些全部都需要CAD建模图纸化，在工厂切割加工后运送到现场。但是，工业生产加工并不都是完美的，建筑最难的地方也在于此，因为建筑是一件单件成形的产品。其实，即使是在"上海保利大剧院"现场，最初也因为产品的大小不匹配，导致那些重要的连接位置无法收口，也是失败连连。

可是，现场施工队伍每次在遇到这种情况时，都会制作现场大样，用实际尺寸进行研究，不断重复改进，一步一个脚印地向前迈进，也正是由于他们的坚韧与努力，才让这座建筑的完成度超越了设计师的想象。

## 在实现建筑的过程中

通过"上海保利大剧院"，我再次深切体会到，建筑的本质并不在图板上，而在现场，它蕴藏在整个建设过程中。

在我们日常生活中的建筑活动，其现实往往遭受时间与经济的制约。在这样的前提下，建筑师选择这种需要通过一个接一个地解决问题才能向前推进的困难过程，似乎从某种意义上来说是很不合理的。在建筑设计界，如今的主流也是要排除粗俗之气，并向着凝练与抽象化的方向发展。

话虽如此，不管形势如何转变，想要创造建筑的是人类，实际使用建筑的也是人类。建筑设计为了适应我们这些充满矛盾且不完美的人类的需要，有时真的需要过度地较真那些不合理的敏感度，以及人类温暖的肉体感知和身体的感官思考。正是在这种不断反复的自我斗争纠缠中，才能培养出空间的紧张感，在流逝的时光中才会愈发深刻地感受到创作的快乐。这种快乐是真真切切存在的。

当我在亚洲的其他国家开展工作时，常常因被不同制度、管理体制所笼罩的建筑现状而感到困惑，有时也会迷失了方向，但也正是这种发展过程所激发出来的新鲜的挑战精神，才让"上海保利大剧院"的实现成为了可能。我由衷地觉得，这个项目启发了我迈向未来建筑的全新可能性，同时也为自己能成为这个项目的设计师而深感幸运。

[安藤忠雄　建筑家/上海保利大剧院主创设计师]

P25: 2006 Tadao Ando's exhibition in Shanghai
P27: Tadao Ando came to the construction site to direct work

25页：2006年安藤忠雄上海建筑展
27页：安藤忠雄亲临施工现场指导工作

CHAPTER 1 | 第 1 章

# Design Process
## 方案演绎

**Shanghai**
上海

**Jiading**
嘉定

**Jiading New City**
嘉定新城

**Site**
基地

**Site Analysis: Theatre**
剧院分析

**Program**
项目内容

**Preliminary Study Process**
初期探讨

**Plans**
方案评述

- Plan A / 方案 A
- Plan B / 方案 B
- Plan C / 方案 C

**Design Concept**
设计理念

**Formal Concept**
形体概念

**Spatial Composition**
空间构成

**Design Concept**
设计方案

- Entrance Hall / 入口大厅
- Auditorium / 观众厅
- Visual Line Analysis / 视线分析
- Traffic Streamline, Fire Safety, and Greenery Analysis / 交通流线、消防、绿化分析
- Landscape Analysis / 景观分析

**Drawings**
图纸

# Shanghai
上海

In terms of population, Shanghai is the largest city in China and one of the largest metropolitan areas in the world with over twenty four million inhabitants. It is located on China's mid-eastern coast at the mouth of the Yangtze River.

Originally a fishing and textiles town, Shanghai grew to importance in the middle of the 19th century due to its favorable port location, and was one of the cities opened to foreign trade by the Treaty of Nanking in 1842. The city flourished as a centre of commerce between East and West, and by the 1930s had developed into a multinational hub of business and finance. Economic reforms in 1990 resulted in intense development and financing in Shanghai, and in 2005 it became the world's largest cargo port.

The city is an emerging tourist destination renowned for its historical landmarks, such as the Bund and Xintiandi. And in Pudong, Shanghai's modern and constantly expanding district, notable landmarks like the Oriental Pearl Radio & TV Tower and the Shanghai Tower grace the city skyline. The new direction of modern urban development makes Shanghai a cosmopolitan centre of culture and economic development. Today, Shanghai is the largest centre of commerce and finance in Mainland China, and has been described as the "showpiece" of the world's fastest growing economy. There is now a strong focus by city planners to develop more "public green spaces" to improve the standard of living for Shanghai's residents, echoing the "Better City-Better Life" slogan of the 2010 Shanghai Expo.

上海位于中国东部沿海的长江口位置，是中国最大的城市，人口超过 2,400 万，其市区规模在世界上也是屈指可数。

上海最早是以盐、渔、棉为主要产业的城镇，其转变的重要转折点在 19 世纪中叶，由于良好的港口位置，1842 年签订的《南京条约》使其作为通商口岸正式开埠。1930 年，上海凭借其作为东、西方商业关系的中心地位，成为跨国公司聚集的金融和商业中心。而 1990 年的经济改革更促进了上海新一轮的发展，直至 2005 年，上海已经成为世界上最大的货运港口。

上海也是一个新兴的旅游胜地，拥有诸如外滩、新天地等著名的都市风景。在现代化不断扩大的浦东新区也拥有了如东方明珠广播电视塔和上海中心大厦这样极具特色的城市地标。新的城市发展方向要求上海成为一个以文化和经济发展为中心的国际大都会。今天，上海已是中国最大的商业和金融中心，也成为了世界上经济增长最快的"精品都市"。以"城市，让生活更美好"为主题的上海 2010 年世博会成为了都市现代化、人性化规划改造的动力，上海将建设更多的"公共绿地"，来提高居民的生活质量。

# Jiading
嘉定

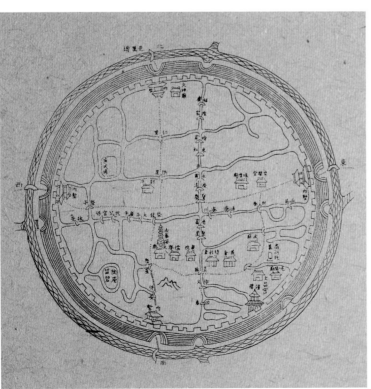

The land on which Jiading is situated was formed 7,000 years ago, while the town itself was established 1217 CE. During the Qianlong period of the Qing dynasty, Jiading gradually developed into an important town of the lower Yangtze region, and has preserved the rich historical and cultural heritage of China's Jiangnan region.

Jiading District lies in Shanghai Municipality's northwestern part, at the heart of the area that is the developing international auto-industry sector. Its meridian coordinates are 121° 15'E and 31° 23'N, and its total area is approximately 463.55km$^2$. The entire area of Jiading is flat, has a water surface ratio of 7.71%, and an average river density of 4.12km/km$^2$. Jiading District has a solid industrial base. By the end of 2014, the value of its grossing industries reached 508.27 billion yuan, and its total profits reached 25.77 billion yuan. Currently, its economic development plan hinges heavily on the automobile industry, in addition to the information technology and other high-tech industries.

There are many places of historic interest and scenic beauty in Jiading's beautiful landscape. Classic examples include Zhou Bridge and Fahua Tower, both of which were built in the Song dynasty. The overall architecture in Jiading is classical, a style that is reflected in its bridges, canals, and residential streets. In Jiangnan, Jiading is known as the "town of education," which, given the Chinese reverence for education, is one of the highest honors that can be bestowed upon a city. In the south of Jiading is the splendid and classic Confucian Temple, built in the twelfth year of the Jiading period during the Song dynasty. On the eastern side of the Confucian Temple is the Danghu Academy, the sole Qing academic institution that has survived in Jiading to this day. Opposite the Confucian Temple is Huilongtan Park, surrounded by greenery and water. It is a leisurely tourist destination in the town. In the east of Jiading, there is also Qiuxia Garden, one of the five famous classical gardens in Jiangnan.

嘉定所在的陆地形成于约7,000年前，立县于公元1217年。清乾隆年间，嘉定逐步成为长江下游的重要城镇，是名副其实的江南历史文化名城。

嘉定区位于上海市的西北部，是建设中的上海国际汽车城所在地。其中心位置在东经121°15'，北纬31°23'，总面积达463.55km$^2$，全境地势平坦，全区水面率7.71%，河道总长1,800余公里，平均河网密度为4.12km/km$^2$。

嘉定区工业基础雄厚，至2014年年末，实现规模以上工业总产值5,082.7亿元，实现利润257.7亿元，形成了以汽车零部件为特强产业，电子信息等高科技产业为支柱的工业经济发展新格局。

嘉定区水秀地灵，名胜古迹众多。嘉定城中的法华塔、州桥皆建于宋代，其周围的小桥流水、民居街巷犹不失古镇风韵；嘉定自古崇尚教育，"教化之地"的美称闻名江南；南城的孔庙建于宋嘉定十二年，殿堂门庑，高壮华好；孔庙东侧的当湖书院是沪上仅存的清代书院建筑；与孔庙一水之隔的汇龙潭公园，绿水环抱，是城内又一休闲观光点，而东城的秋霞圃更是蜚声江南的上海五大古典园林之一。

P30: Top, Sketches of Lujiazui and Shanghai Oriental Art Centre from Tadao Ando; Bottom left, Night views of the Bund and Xintiandi; Bottom right, View overlooking Lujiazui Financial District
P31: Top, Regional map of Shanghai, the red part is Jiading District; Bottom, Map of Old Jiading City

30页：上，安藤忠雄笔下的陆家嘴和东方艺术中心；左下，上海外滩和新天地夜景；右下，远眺陆家嘴金融中心
31页：上，上海市区域图，红色部分为嘉定区；下，嘉定古城图

# Jiading New City
## 嘉定新城

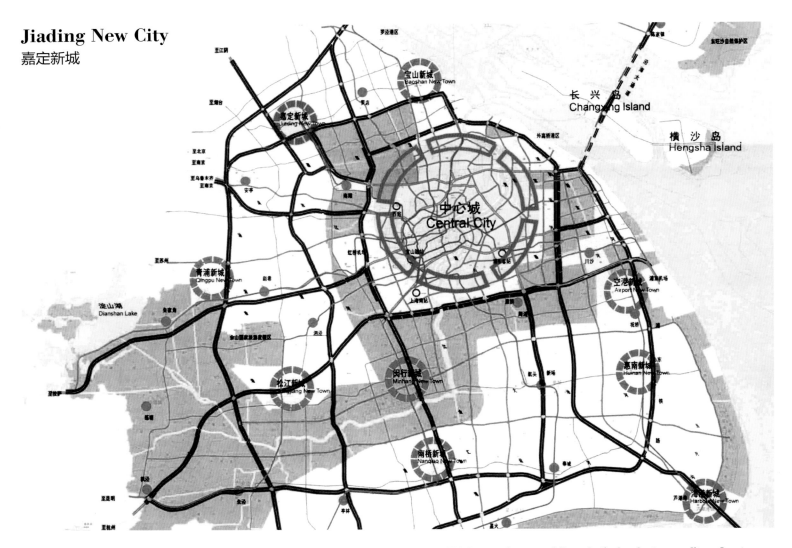

P32: Map of Shanghai Central City and its surrounding satellite cities
P33: Top, Plan of the city centre in Jiading New City; Bottom, Photo of a model of the city centre of Jiading New City. The body of water is Yuanxiang Lake. Located beside the lake are the Shanghai Poly Grand Theatre and Commercial & Cultural Centre

32 页：上海市中心城区与周边卫星城分布图
33 页：上，嘉定新城中心区规划图；下，嘉定新城中心区模型照片，水池部分为远香湖，湖畔为上海保利大剧院和商业文化中心

Jiading New City is located to the south of the Old City, and is one of Shanghai's developing satellites. Its city centre is bounded by Forest Road in the north, Liuxiang Road in the east, Wenzao River in the south, and North Jiasong Road in the west. The total land area is about 122km$^2$. It is projected that, by 2020, the population for Jiading New City's main urban areas will be 500,000 people, and that the land area used for construction will reach 66.8km$^2$.

The city's core industries will consist of the following: a modern service sector, a world-class sports and leisure sector, and high-tech industries. The city plans to become the regional core of Shanghai's suburbs, with its unique cultural and civic characteristics and its creative focus. At the same time, the suburban areas will aim to define themselves as the heart of Jiading New City, which in turn will strive to become the model for strategic development in Shanghai's satellite areas.

The northern part of the city will be the main development area. The east will be developed as a mixed-use area. The city centre will serve as the administrative and civic centre, while the residential areas will be allocated to the west and to the city's periphery. The major green spaces will include the Ziqidonglai Green Belt, Yuanxiang Lake, and other green belts lining the city's canals.

嘉定新城位于嘉定老城区的南部，是上海郊区近期重点发展建设的新城之一。主要规划范围北至森林大道、南至蕰藻浜、东至浏翔公路、西至嘉松北路，总用地面积约 122km$^2$。到 2020 年新城主城区实际居住人口将发展到 50 万人，城市建设用地将要达到 66.8km$^2$。

新城将以现代化服务业、世界级体育休闲产业和高科技产业为核心，打造具有独特人文魅力和城市特色、强大的集聚力和持续的创新力的上海都市圈区域性核心城区，同时成为上海郊区发展战略重点的示范城区以及嘉定组合式新城的主体与核心。

嘉定新城开发主要集中在新城的北部地区，东部为混合性用地，公共建筑用地居中，西边及其周边为居住用地，主要绿地包括紫气东来景观轴线、远香湖和沿运河的带状绿化空间。

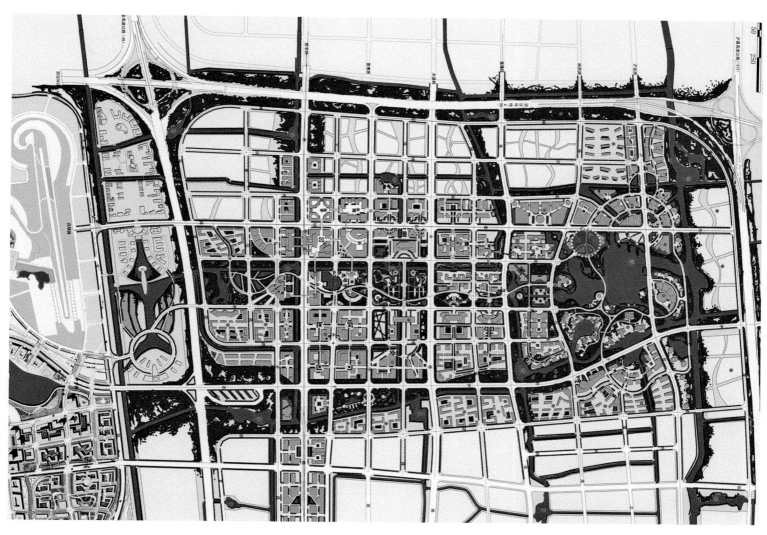

# Site
基地

PP34-35: Aerial view of the Shanghai Poly Grand Theatre's construction site
PP36-37: Left, Site location; Top right, Aerial photos of the site before and after construction
P38: Top, Tadao Ando's sketch of the Sydney Opera House; Bottom (from left to right), The respective interiors of the Paris Opera House, Sydney Opera House, Copenhagen Opera House, Oslo Opera House, Paris Opera, Suntory Hall, and Opera Lyon

34-35 页：上海保利大剧院基地鸟瞰图
36-37 页：左，基地位置图，右上，基地周边整治前和开发后的航拍照片
38 页：上，安藤忠雄手绘悉尼歌剧院；下，由左至右依次为：巴黎歌剧院室内场景、悉尼歌剧院、哥本哈根歌剧院、奥斯陆歌剧院、巴黎歌剧院、三得利音乐厅、里昂歌剧院

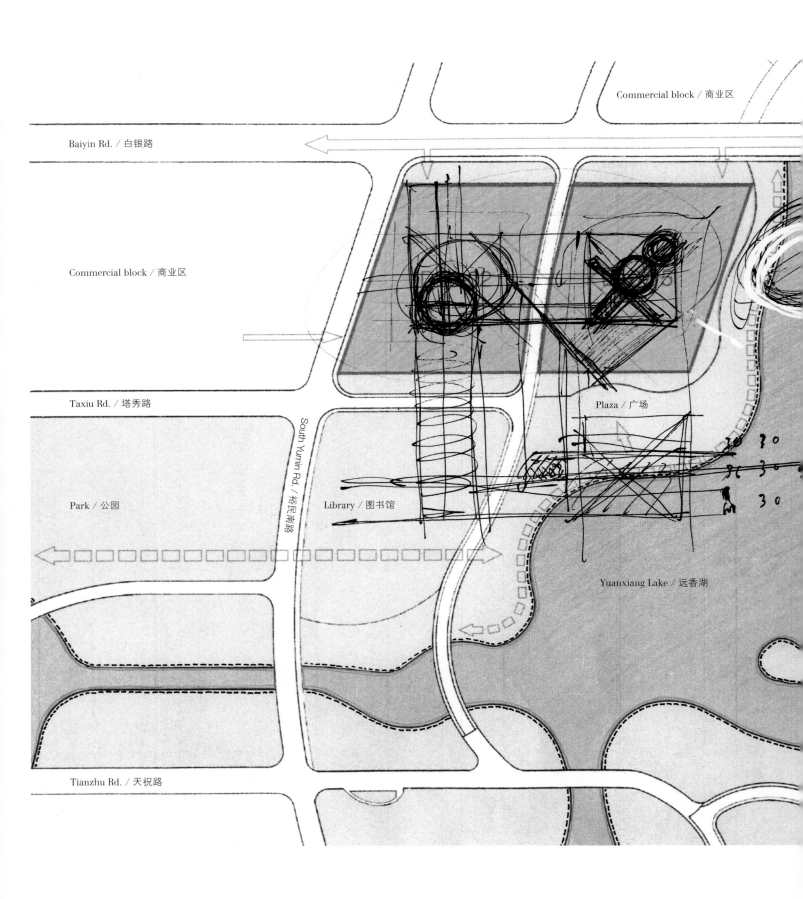

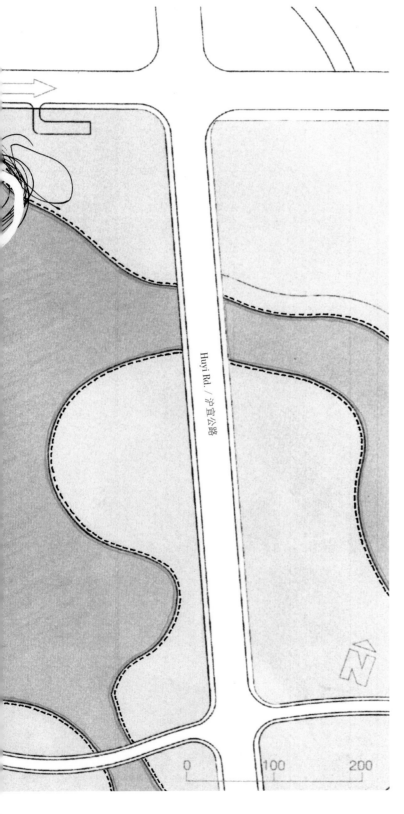

Our site is located in the southeast of Jiading New City, at the corner of the administrative and cultural centre's park, bordering Yuanxiang Lake. It consists of two city blocks, which includes the Shanghai Poly Grand Theatre facing Yuanxiang Lake and the Poly Commercial & Cultural Centre.
At the site's northern edge, across Baiyin Road, lies a lively commercial centre, while to the west there is a large business district. A large natural area, consisting mostly of greenery and water, is located at the south, giving the opera house complex an outstanding view: to the southwest, a park; to the south, a municipal library; and to the southeast, Yuanxiang Lake. Facing the lake will be a plaza and promenade, which will function as civic spaces of public discourse and recreation.

基地位于嘉定新城东南侧，在行政文化中心内的公园一角，紧邻远香湖畔。基地分为保利大剧院和商业文化中心两个区域，其中保利大剧院坐落于面向远香湖的地块上。

与基地北侧相隔一条白银路的是嘉定新城的商贸中心，西侧面向商业街区。基地南侧，以水和绿化为主的自然风景缓缓铺开，将西南侧的公园、南侧的图书馆还有东南侧的远香湖美景尽收眼底，并计划在临湖的位置设置广场和散步道，使其成为市民休息交流的场所。

# Site Analysis: Theatre
剧院分析

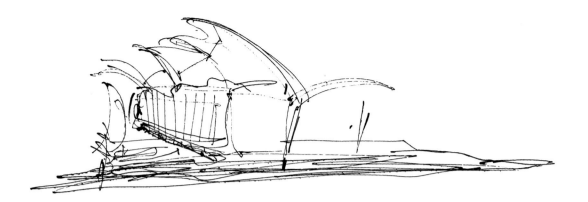

Since the completion of the Paris Opera House in the 19th century, the Theatre has not only become the stage of actors, but also the special stage for spectators to display themselves. Here, in the space's grand opulence, fantastic stories emerge to break the mundaneness of everyday life.

The Theatre is a place to inspire the passions of its visitors. The entrance hall and the foyer are the "stages" of the citizens, who come to escape from the constraints of everyday life, and immerse themselves in a magnificent spatial experience. It is from this convergence of the spiritual, musical, and operatic experiences that ignites people's passion for art.

Moreover, the Theatre is not only the face of the city, but is at the core of its iconic existence.

Shanghai, as a city welcoming tourists from all over the world, needs a building that will live forever in their memories, as well as a unique landmark that is fully integrated into the natural scenery surrounding Jiading New City.

自19世纪巴黎歌剧院落成以来，剧院不但成为演员的舞台，更是成为了观众们展现自我的特殊舞台。在这里各种经典故事层出不穷，展现了其非日常性的、豪华、盛大的空间场所特征。

剧院是让来到这里的人们胸中涌动激情的场所。入口大厅和休息厅是市民的"舞台"，市民来到这里可以逃离日常生活的束缚，迎来一种盛大华丽的空间体验。然后，以这样高昂的心情与音乐、歌剧相撞，迸发出激越的艺术火花。

另外，剧院不仅是都市的"容颜"，更是都市标志性的存在。

迎来世界各国观光客的上海，需要一座可以常驻在人们记忆中的建筑；同样也需要一座与环绕嘉定新城的美丽自然风景相融合，并且拥有其独自个性的地标性建筑。

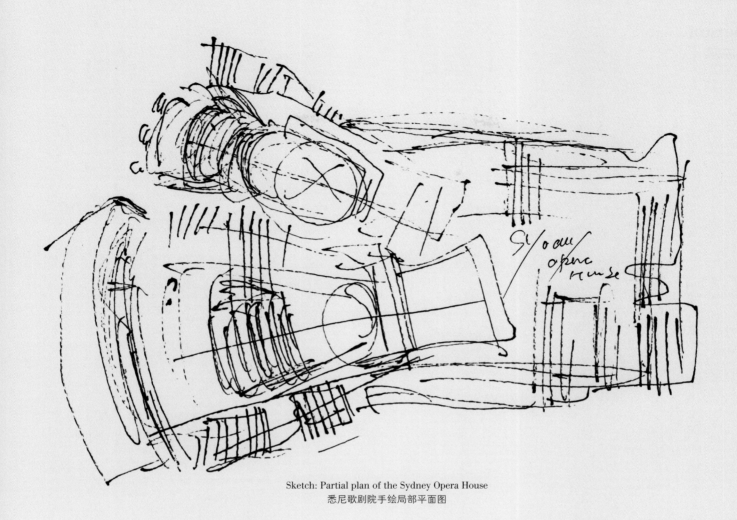

Sketch: Partial plan of the Sydney Opera House
悉尼歌剧院手绘局部平面图

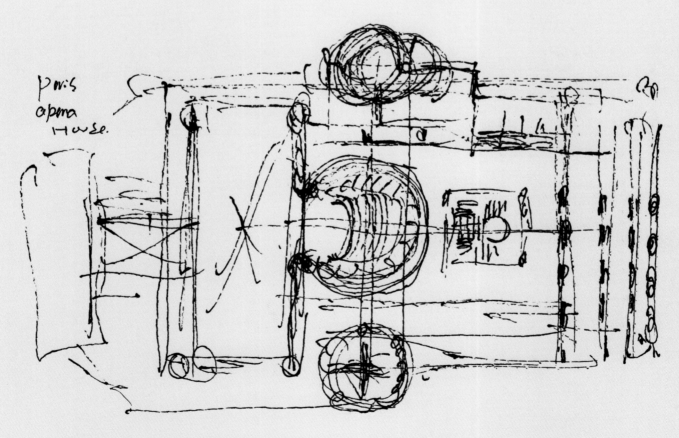

Sketch: Plan of the auditorium of the Paris Opera House
巴黎歌剧院手绘观众厅平面图

# Program
项目内容

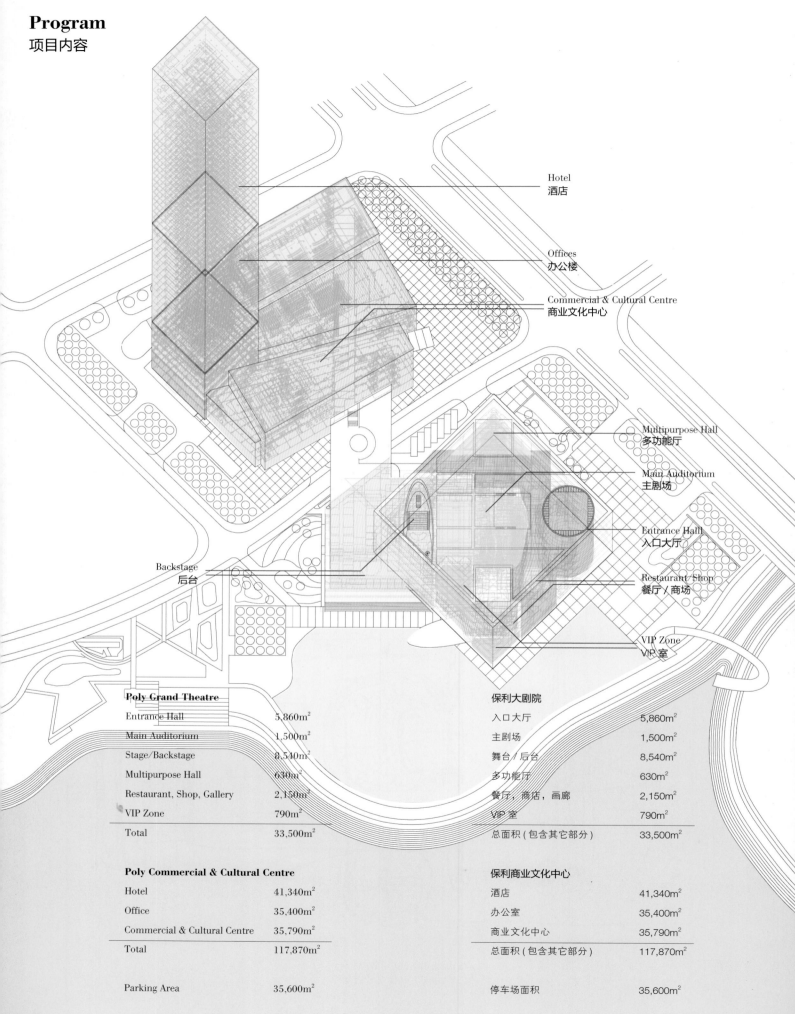

Hotel 酒店
Offices 办公楼
Commercial & Cultural Centre 商业文化中心
Multipurpose Hall 多功能厅
Main Auditorium 主剧场
Entrance Hall 入口大厅
Restaurant/Shop 餐厅/商场
VIP Zone VIP室
Backstage 后台

**Poly Grand Theatre** / 保利大剧院

| | | |
|---|---|---|
| Entrance Hall / 入口大厅 | 5,860m² | 5,860m² |
| Main Auditorium / 主剧场 | 1,500m² | 1,500m² |
| Stage/Backstage / 舞台/后台 | 8,540m² | 8,540m² |
| Multipurpose Hall / 多功能厅 | 630m² | 630m² |
| Restaurant, Shop, Gallery / 餐厅,商店,画廊 | 2,150m² | 2,150m² |
| VIP Zone / VIP室 | 790m² | 790m² |
| Total / 总面积(包含其它部分) | 33,500m² | 33,500m² |

**Poly Commercial & Cultural Centre** / 保利商业文化中心

| | | |
|---|---|---|
| Hotel / 酒店 | 41,340m² | 41,340m² |
| Office / 办公室 | 35,400m² | 35,400m² |
| Commercial & Cultural Centre / 商业文化中心 | 35,790m² | 35,790m² |
| Total / 总面积(包含其它部分) | 117,870m² | 117,870m² |
| Parking Area / 停车场面积 | 35,600m² | 35,600m² |

# Lighting up the City
建筑点亮城市

Interview with SUN Jiwei
采访孙继伟

Interviewer: MA Weidong / 采访者：马卫东

**Q:** It has been said that there had been a design that preceded Mr. Ando's for the Poly Grand Theatre. What did you think of the previous design?

**A:** When the government entrusted the Theatre project to Poly Property, a Chinese enterprise specializing in Theatre construction and management, the latter invited a domestic design company to submit a design. That design was also quite good. However, in the context of strategic development in Jiading New City, such a landmark is bound to become an important urban development opportunity, bringing a certain amount of visibility and influence to Jiading. Although we did not envision it to be like the Sydney Opera House, we wanted it to be a building that can light up the city and the region. We also wanted it to be a positive force in the region's urban landscape and culture. Therefore, the earlier design was unsatisfactory with respect to increasing Jiading's cultural visibility and influence.
Jiading New City's initial development strategy was to commission of group of Chinese and foreign master architects to design several buildings in the city, thereby illuminating the city with a series of notable buildings, such as Jiading Library and the Youth Centre. The Poly Grand Theatre is undoubtedly the most important cultural landmark, hence the need to invest heavily into the project, and the need to find a world-class architect to design it. After discussing the matter with Poly Property, the enterprise also became aware of the building's significance.

**Q:** As the chief administrator of Jiading District at the time, what were your thoughts on Poly Grand Theatre's proposed purpose in Jiading New City?

**A:** First, in planning a new medium-sized city as an important hub for the Yangtze River Delta Region, it is important to include a cultural landmark in the planning. Second, once the Theatre is completed, it would not only serve Jiading New City, but also the entire surrounding region, including Kunshan, Taicang, and Suzhou. As long as it is sufficiently attractive and well-managed, it will become one of the region's great cultural spaces. We believe that the Theatre's high activity level and adherence to high standards bode well for its future. As for the Poly Grand Theatre itself, its design and location makes it the finishing touch to Yuanxiang Lake. Although many buildings around the lake are designed by famous architects, like the Shanghai Jiading Public Library, which is known for its aesthetic beauty, I feel that the Poly Grand Theatre will become even more important than its predecessors, since it will become the focal point for the development of the entire district of Jiading. In many other countries, buildings attract visitors, for they are the masterpieces of great architects. The Sydney Opera House and the Paris Opera are two examples. A city becomes the cultural centre of an entire region by virtue of its cultural landmarks.

**Q:** How did you decide on Tadao Ando as the project's architect? We know that Ando is a world-famous architect, especially in the design of art galleries and museums. However, he had never designed any well-known Theatre spaces at that time.

**A:** In fact, we had many candidates at the time. However, we had to take into account of each architect's schedule, design costs, and many other issues. In the end, we were pleased to select Tadao Ando, as he is a world-famous architect. Another important reason for choosing Mr. Ando was that he has a basic understanding of the Chinese construction industry. Had we hastily chosen a world-famous architect who did not know much about China, the design would not have worked, especially if its material, technological, and procedural standards far exceeded China's current standards; we would not have been able to construct it, much less afford it. Therefore, once we considered all of the factors, we selected Mr. Ando. We would say that Mr. Ando selected us as well. Mr. Ando has never designed a large Theatre space before, but his understanding of cultural landmarks and grasp of space are very unique. If he were to design the Theatre, he would give us a very special design based on his aesthetic style. This is indeed what happened. I was worried about the possibility of needing to reconcile Ando's design with our vision of the project—what if the difference was so great that no resolution was possible? Maybe, this architect is not suitable for the project. Or maybe a completely unachievable design is presented. Over thirty models were submitted during the first round of design submissions for the Poly Grand Theatre, but all of them were unsatisfactory. At that time, we were really worried. However, during the second round of submissions, when that rectangular model was presented before us, I immediately thought, "that's it." It really was a wonderful design: its interpretation of the concept (of the Theatre) and its ideas for its interior space were magnificent.

**Q:** We know that you were really concerned with the project, involving yourself in its planning and design, as well as personally guiding the work. In the early stages of the project, Mr. Ando proposed the "cultural kaleidoscope" as the design concept. What is your opinion and understanding of the design?

SUN Jiwei
孙继伟

Master of Architecture, Tongji University
Doctor of Engineering
Current Deputy Secretary General of the Shanghai Municipal Government
Deputy Secretary of district committee(of the CPC),
Secretary of district government(of the CPC) of Shanghai Pudong New Area,
Mayor of Shanghai Pudong New Area

同济大学建筑系建筑设计专业硕士，工学博士
现任上海市政府副秘书长
中共浦东新区区委副书记，区政府党组书记、区长

问：据说在安藤先生的设计方案之前，保利大剧院已经有了一个设计方案。在此，想请您谈一谈，您是如何看待之前的那个方案的？

答：政府将大剧院的项目委托给保利置业的时候，保利置业作为在国内专业建设和运营剧院的企业，在接受委托后便积极地邀请国内一家设计公司设计了一个方案，其实那个方案设计得也很好。但是从嘉定新城未来发展的策略和战略角度来考虑，作为一座具有地标性意义的文化建筑，势必会成为一座城市发展的重要契机，并为嘉定新城带来一定的知名度和影响力。虽然不一定能像悉尼歌剧院般，一座建筑带动一座城市，但至少希望通过一座重要的文化设施建筑来点亮一座城市、点亮一个地区，为这个地区的城市景观和城市文化起到一种正向的作用。因此对于之前的方案设计，就其知名度和影响力来说还是不够的。

嘉定新城最初的发展战略是希望由一批建筑大师或明星建筑师，在嘉定新城建造若干建筑，通过群星的光芒来照亮整座嘉定新城，例如嘉定图书馆、青少年活动中心等建筑都是由国内外知名建筑师来设计的。而保利大剧院无疑是最重要的文化设施建筑，无论是对它的投资，还是它的地理区位都占有很重要的地位，因此希望由一位世界级的建筑师来进行设计。在与保利置业商量后，他们也意识到了这一想法的重要性。

问：作为当时的嘉定区政府管理者，您是如何看待保利大剧院在嘉定新城所处的地位和作用？

答：在嘉定新城建造一座大剧院，一方面，由于新城在规划发展成为中等城市乃至长三角地区重要的节点城市时，本身就需要有一座地标性文化建筑；另一方面，当时计划这座剧院建成后，不仅仅只是为嘉定新城服务，更是为整个区域服务，其中包括昆山市、太仓市、苏州市等许多周边城市，只要这座剧院有足够吸引力，且经营力度足够大，它就会成为整个区域内

的剧院，所以它对于整个区域来讲也是非常重要的。我们希望这个剧院能具备一定的能值和水准，这也有利于今后的经营。

从保利大剧院建筑本身来说，无论是景观，抑或是地理位置，都使它成为远香湖的点睛之笔，虽然远香湖周边已经有许多知名建筑师设计的建筑，例如被誉为环境最优美的图书馆——上海嘉定区图书馆也是一座不错的建筑，但我觉得保利大剧院的重要性尤为突出，它将成为整个嘉定新城发展过程中的一个激发点，通过建筑的吸引力来增加嘉定新城的魅力。在国外，许多建筑由于是大师的杰作，所以吸引了更多的人前来参观，诸如悉尼歌剧院、巴黎歌剧院一样，慕名而来参观建筑的人络绎不绝，正是由于建筑的吸引力使一座城市成为整个区域的文化兴奋点和焦点。

问：您是如何决定邀请安藤忠雄先生来担任这个项目的设计师呢？我们知道安藤先生是世界著名的建筑大师，尤其是在美术馆及博物馆之类的文化建筑领域独树一帜，可是在当时他并没有特别出名的大剧院设计实例。

答：当时，其实有多位大师的人选，而与一位大师的合作，需要协调包括设计师的档期以及设计费等诸多问题，最终我们很荣幸地邀请到了安藤忠雄先生，而他的知名度也足以承担这个项目。此外，邀请安藤先生还有一个重要的原因，就是他对中国的施工状况以及造价标准都是有所了解的，如果贸然地选择了一位不熟悉中国的世界级建筑师，而他对材料、技术、工艺的要求都非常高，甚至超出了我们的造价标准，则是我们所不能承受的，即使设计出来，也未必能够造出来。因此综合考虑了所有因素后，我们选择了安藤先生，同样安藤先生也选择了我们。

安藤先生虽然没有做过大型的剧院建筑项目，但他对文化建筑的理解以及空间趣味性的把握，很有他自己的特色，如果让他设计这座剧院，根据他的设计风格，以及审美角度的不同，也许能设计出一个很特别的方案，事实也的确如我们所想的一样。我曾经也担忧过，请大师的难点就是，如果设计的结果与我们设想的有所差距时，我们如何解决？也有可能这位大师并不适合这个项目，或者最终呈现的是一个完全无法实现的方案。在保利大剧院第一轮方案汇报时，一共做了三十几个模型，都不如我们所想的那样，当时其实还是有些担心的，但在第二轮方案汇报时，当那个长方体建筑模型展现在我们面前时，我的第一眼感觉"就是它了"，无论从理念的解释，内部空间的一些设想，都是一个很精彩的设计。

问：从项目企划到方案确定，我们知道您一直非常关心，并亲自指导工作；在方案初期，安藤先生提出"文化万花筒"设计理念，您是如何看待和理解这一设计构思的？

答："文化万花筒"是一个很精彩的理念——利用几个简单的圆筒穿插于一个纯粹的长方体里，从而创造出生动丰富的空间。在模型讲解时，特别是横向穿插在里面的几个圆筒，构成了半室内半室外的空间，十分有趣；还有一处就是入口大厅，它是一个垂直穿插于长方体内部的圆筒空间，让我印象非常深刻的是安藤先生当时讲的一句话："在剧院开场之前，入口

**A:** "Cultural kaleidoscope" is a wonderful concept. Several simple cylinders slice through a pure rectangle, creating spaces that are open and vibrant. During the conceptual interpretation phase, I found some of the intersecting cylinders forming spaces that were both indoors and outdoors to be very interesting. Another one that caught my eye was the entrance hall, in which the cylinder intersects vertically with the rectangle. Mr. Ando said at that time, "Before the theatrical performance, the entrance hall is the audience's performance space; it is the audience's stage," which left a deep impression on me. Mr. Ando paid special attention to the space welcoming the audience, giving it a thick and heavy color. After the project was completed, the space indeed became the audience's stage, so his concept was flawlessly realized. When people enter this space, they will certainly find the experience to be very interesting.

**Q:** On December 22nd, 2014, Mr. Ando held a lecture in the Poly Grand Theatre; before that, he visited the site with you. At that time, the two of you discussed the civic nature of the Theatre. Mr. Ando said that the Theatre's public spaces would still be open to the public even if there was no performance being held, thereby functioning as a public living room of Jiading District, a public space that citizens in Jiading can be proud of. What are your thoughts on its civic nature?

**A:** In my opinion, if we understand the Theatre as a place where performances can be seen, we only grasp its basic function. As a significant cultural landmark of Jiading New City, it is not only a Theatre—it is a work of art. It will attract visitors looking to sightsee and participate in various activities. Its civic nature is obvious. In such a public facility, with half its space on the outside and half of it inside, people are their most vibrant selves in the kaleidoscope. During the design phase, Mr. Ando reserved a public area to hold various civic and cultural activities. Many stories will be written here. Many people will come to visit and experience the space. The space will be open and public.

**Q:** When entrusting the project to Mr. Ando, the client wanted the Poly Grand Theatre to not only become a cultural landmark in Shanghai and China, but a world-class venue comparable to the Paris Opera and Sydney Opera House. What are your thoughts on the Poly Grand Theatre's chances of reaching this lofty goal?

**A:** When we talked about finding a world-class architect for this project, we wanted a building that be world famous by virtue of the architect's popularity, which is why we picked Mr. Ando. From the very beginning of the project, there was a lot of pressure on Mr. Ando. We believe that the project's ultimate success was due to his persistence and the

大厅是观众的演出空间,是一个观众的舞台。"安藤先生特别重视观众入场前所在场所的空间设计,尤其的浓墨重彩,建成后,这些场所也确实成为了观众的舞台,理念实现得非常好。当市民穿梭在这些空间里,也一定会感觉非常有趣。

问:2014年12月22日,安藤先生在保利大剧院举办了讲演会,在此之前您和安藤先生共同参观了保利大剧院,当时您二位有过关于大剧院公共性的一番讨论。记得安藤先生说,大剧院的公共空间在没有演出的时间段,仍然对外开放,对市民开放,使得大剧院承载着嘉定区公共客厅的功能,让嘉定区的市民有一个值得自豪的公共活动场所。那您是如何看待大剧院的公共性?

答:我觉得把剧院的公共性只理解为公众来看一场演出,那还仅仅只是体现了剧院的功能性,当保利大剧院能够成为嘉定新城一个重要的景观和标志性文化建筑时,它就不仅仅只是一座剧院,更是一件艺术品,它本身就能吸引人们前来参观和活动,这种公共性是显而易见的。而这些具有公共性的场所——半室内半室外的空间——穿插在其中的万花筒中最生动的就是人,也就是那些前来参观活动的市民。安藤先生在设计方案的过程中,为这些公共空间留下了足够多的余地,让市民可以在这些空间内举行各种活动,可以发生很多故事,还能吸引很多人来这里参观和体验,由于空间本身的吸引力,必然使它成为一个公共性的空间和一个公共场所。

问:当时在委托安藤先生做设计时,业主方提出的要求是,希望保利大剧院能够成为如同悉尼歌剧院、巴黎歌剧院一样屹立在世界舞台上的一流大剧院,能够作为上海乃至中国的文化建筑新地标。您觉得保利大剧院最终达到这样的目标了吗?

答:一开始在商谈要找世界级大师做这个项目时,就希望能借大师的知名度创造一座具有同样世界级影响力的建筑,因此对于安藤先生也就寄予了这样的希望,这也让他一开始就承受着很大的压力,可以说最后的成功,与安藤先生不断地努力,以及整个事务所停下所有的工作进行专项负责,是分不开的。在与安藤先生的通信中,我感受到了一位建筑师质朴的匠心和对建筑的热忱,深受感动。随着这个项目的一步步实施,保利大剧院的形象也慢慢展现在我们的眼前,同时也离我们当时的目标越来越近。正如我当初看到模型时一样,设计已经具备了一流水平,蓝图也已经有了,接下来就是如何完美地实现;从设计到施工,再到室内,最终呈现的效果其实已经达到了我的期望,我是非常满意的。如果造价能够再充裕些,也许还能建得更加堂皇、更加到位。而剧院最终的音质也受到了普遍的好评,为大家共同的努力给出了最好的回报。

此外,这个项目的成功,一方面来自于政府的支持以及所寄予的希望,另一方面保利置业也投入相当大的心血,并且选择了一支优秀的施工队伍,当然这支队伍能够如此出色地完成该项目,也离不开多方的支持。我记得当时还专门组织了施工队去日本考察学习,安藤先生也专门请了日本的施工队伍,对施工的方式方法进行讲解辅导,并且还推荐了一系列做清水混凝土的工具以及一些施工技巧。当时施工队回来后,也进行了研发和改进,并形成了自己的一套混凝土施工方法,成为国内清水混凝土做得最好的一支施工队伍,上海徐汇区的龙美术馆也是这支队伍完成的,清水混凝土同样完成得非常出色,模板做得更是犹如艺术品一样。

问:我们知道您是建筑专业博士毕业,作为国内少有的建筑专家型官员,其实早在大约十年前,在青浦区,您就和安藤先生有过合作。同时,您和全世界各国的大师级建筑师都有很多合作的项目,在此想请您谈谈,这次和安藤先生的合作,有什么特别不同之处吗?

答:安藤先生本身是一位很容易沟通的设计师,也没有世界级大师的那种傲慢或者固执,他很能从业主需求的角度来考虑问题,属于刚柔相济型。对于我们提出的尤其是使用及品质方面的意见,他还是尽可能地满足;但对于本质的一些艺术创作意图,或是一些精彩设计理念,他都会牢牢坚守。与其他建筑师相比,由于他在世界各地做过许多项目,已经形成了自己独有的一套风格和特色,包括一些设计思想和理念。有些设计师还在不断摸索和积累的过程中,虽然有很好的创造力,但还是处于发展阶段,而这就是一位成熟建筑师和年轻建筑师之间的差异。

问:您任职的青浦区、嘉定区、徐汇区这三个区里,都留下了很多很好的城市开发成果和建筑作品[1],我想保利大剧院也是其中之一。在您的开发理念和思想中,"文化艺术"一直是一个非常重要的关键词,青浦的

---

1  Mayor SUN Jiwei once served, on separate occasions, as the Deputy Mayor of Qingpu District, the Mayor of Jiading District, and the Secretary of Xuhui District Committee. During his terms in office in these three districts, he guided the urban development of Qingpu New City, Jiading New City, and Xuhui Riverside (West Bund), leaving behind many excellent buildings which have received high praise from the industry. Qingpu New City mainly consists of the Qingpu New City Construction Management Centre, Qingpu Xiayu Kindergarten, The Green Pine Garden, Qingpu Thumb Island, etc. Jiading New City includes Jiading Urban Planning Exhibition Centre, Jiading New City Kindergarten, Jiading New City Library & Cultural Centre, and the Shanghai Poly Grand Theatre. In Xuhui District, there are the Long Museum, De Museum, the West Bank Cultural Corridor, West Bank Media Port, and other cultural facilities.

1  孙继伟区长分别担任过上海市青浦区副区长、嘉定区区长和徐汇区区委书记。在这三个区的任职期间,他主导开发了青浦新城、嘉定新城和徐汇滨江西岸,留下了很多优秀的建筑作品,获得业界的高度评价。青浦新城主要有青浦新城建设管理中心、青浦夏雨幼儿园、青松外苑、青浦夏阳湖浦阳阁等作品;嘉定新城内有嘉定城市规划展示馆、嘉定新城幼儿园、嘉定新城图书馆新馆及文化馆、上海保利大剧院等优秀建筑作品;徐汇区则有龙美术馆、德美术馆、西岸文化走廊、西岸传媒港等文化气息浓厚的建筑作品及公共设施。

孫継偉 様

ご無沙汰しています。
先日 保利劇場の工事現場
を見に行きました。
おかげさまで ここまで進めること
が 出来ました。
思いがけない質の高い仕事が
出来たと感謝しています。
世界に誇れるものを。と言われて
いたのに少しは近づけたのだと
思います。

19/03/2014 Tadao Ando
安藤忠雄

**Mr. SUN Jiwei**
Long time no see. How are you recently?
I went to the construction site of Poly Grand Theatre a few days ago.
Thanks for taking care of this and it made the construction go smoothly.
The quality of the completion is much higher than what I expected. I deeply appreciate what you did.
You used to say you wanted to make a project which will be profusely praised by the whole world. In my opinion, this day is coming.

Tadao Ando 19 March 2014

孙继伟先生：
　　久疏问候，您近来可好？
　　前几天，我去看了保利剧院的施工现场。
　　承蒙关照，至此工程得以顺利地进展。
　　工程完成的质量之高超乎了我的想象，对此我表示深深的感谢。
　　之前您说要做出让全世界都称道的作品来，我想这个日子已经临近了。

安藤忠雄　2014年3月19日

fact that his entire firm stopped working on all other projects to work on the Theatre. In my written exchanges with Mr. Ando, I was deeply moved by the architect's ingenuity and enthusiasm for architecture. As the project unfolded step by step, the image of the Poly Grand Theatre materialized, moving us toward our ultimate goal. Just like the model, the design and the blueprint were of the highest quality. The next step was to execute the design perfectly. And the project in its entirety-from design to the construction process to the interior-exceeded my expectations. If we had a greater budget, the building could have been more grandiose. Its acoustics were widely lauded as well. All of these accomplishments provide a great return for our collaborative efforts. The project's success can also be attributed, on the one hand, to government support and expectations, and on the other hand, to Poly Property's considerable efforts. Poly Property assembled an excellent construction team, as well as many other support teams. I remember the construction team specifically organizing a trip to Japan to learn from the Japanese team the methods and means of constructing fair-faced concrete. The Japanese team also recommended certain machinery to be used in the fair-faced concrete process. After returning to China, the construction team had to develop and improve the process, to create China's own set of fair-faced cornet construction methods. As a result, the team became the best Chinese construction team for fair-faced concrete. It also completed the Long Museum project in Shanghai's Xuhui District-the fair-faced concrete was well-made, and its template was like a work of art.

**Q:** We know that you have a Doctor degree in architecture, and that you had worked with Mr. Ando in Qingpu District ten years ago. Being one of the few government experts on architecture in China, you have worked with many world-class architects. Was there anything special about your collaboration with Mr. Ando?

**A:** It is really easy to communicate with Mr. Ando, who, despite being a world-class architect, is neither arrogant nor stubborn, and is able to consider problems from the client's perspective. He is always friendly and confident. As for our opinions and requests, he would satisfy them as much as possible. However, in order to preserve certain artistic elements or design concepts, Mr. Ando would insist firmly upon them. When compared with other architects, he has his own unique style, along with some interesting design ideas and concepts, having designed so many buildings in the world. Some architects are still trying to figure out their individual style, and that is the difference between an established architect and a young one.

**Q:** Your term in office in the districts of Qingpu, Jiading, and Xuhui has yielded good results in urban development. I think the Poly Grand Theatre can be considered one of them. The idea of "cultural art" has always been important in your views on urban development. The library in Qingpu, the Theatre in Jiading, and the West Bund art corridor have all received high praise from the public and professional sectors. Currently, you have launched a new round of urban projects in Pudong New Area. What are your thoughts on urban development? What is your conception of development?

**A:** With regards to development, we are still learning. During the last round of suburban and urban development in Shanghai, we learned some new things and identified some problems, such as our desire to complete large-scale projects in a short amount of time, resulting in shoddy construction. As a rapidly developing city in China, the urbanization of Shanghai and its suburbs provides a model of sorts for the development of small-and medium-sized cities. However, the essence of development is not in its speed or scale, but in its quality. The key point in development, then, depends on improving the aesthetic charm and functionality of a new city. Now, thanks to our earlier experience, we have a clear concept of urban development. During my time in office in Qingpu and Jiading, the expansion of the city posed many problems: where were the people coming from, who were coming there, and what were its attractions? At the time, we only hoped that the city would stand out from other new cities, and the only way to accomplish that was to ensure that it was well-constructed. It was then that I raised the "four-fame" principle: to have famous buildings be designed by a famous designer and constructed by a famous developer and a famous construction company. Jiading District would then be illuminated by the collective celebrity of its buildings. Without a doubt, a group of good buildings and a good environment alone do not make for a good city, since such a city would lack a cultural essence. I once made an analogy: a good city is just like aged wine. Aged wine isn't just 53% alcohol and 47% water—it needs subtle things. And that is a city's cultural essence, the accumulation of art and culture over time. And this is what I always strive for during my terms in office.

Q: Last question. When you participate in a government-led development project, what kind of dreams are you investing in its architecture?

A: When I gave my speech at Tongji University, I strongly encouraged students majoring in programming and architecture to work for the government after graduation. This would not only provide an influx of professional talent into the city planning sector, but also contribute to the development of the entire construction industry. When the government is guiding urban development, there needs to be a good decision-maker driving the process. I am a government official who is able to realize my personal dream. The current conditions and amount of resources in China allow for many ideas to be put into practice. I think the government plays a more important role than architects working on a certain project or city block. I hope that more people can make a positive contribution to the city.

(Interviewer: MA Weidong  Founder of CA-GROUP /Member of Architecture Society of Shanghai China)

图书馆、嘉定的大剧院、徐汇滨江的西岸艺术走廊，都获得了社会和专业领域内的高度评价。现阶段您在浦东新区又展开了新一轮的城市建设，请问您是如何看待城市建设的？您的开发理念又是什么？

答：我们也是在学习和积累的过程中，在上海前一轮的郊区和新城开发过程中，我们吸收了一些经验，同样也看到一些弊端，比如过度追求速度、规模，造成了一些粗制滥造的结果。上海作为一个走在中国发展前沿的城市，其城镇化的过程和郊区新城的建设，为中国快速推进中小城市发展提供了一些样板和示范，而这些样板和示范不在于速度和规模，更重要的是品质，怎么样能在建设新城过程中，把品质做上去，把新城地区魅力和吸引力做上去，这才是关键。现在大家对新城的概念都已经很清楚了，在我任职青浦和嘉定期间，新城的扩张面临许多问题，比如人从哪里来，谁会来新城，新城的吸引力在哪里，魅力点在哪里。当时我们所希望的就是自己参与建设的新城能够在众多新城中脱颖而出，最佳的方式就是把品质做好。当时我提出了四个"著名"，以著名的设计师、著名的施工单位和著名的开发商打造著名的建筑，让群星的光芒点亮整个嘉定。当然仅仅有一批好的建筑，好的环境，未必能造就一个好的城市，好的城市还需要有一定的文化内涵。我曾作过一个比喻，一个好的城市，就像一壶陈年美酒，仅仅只是53%酒精+47%水，是无法酿成一瓶好酒的，这中间还有许多微妙的东西，这即是城市的文化内涵——文化艺术和时间的累积，这也是我任职期间一直在坚持追求的。

问：最后，想请您谈一谈，通过这些政府主导的开发项目，其实寄托了您怎样的建筑梦想？

答：我去同济大学讲演时，就非常鼓励那些学规划、学建筑的学生毕业后能投身于政府工作，这样既能为整座城市的建设提供更专业的帮助，也能为推动整个建筑业的发展起到很好的作用。在政府主导城市建设的过程中，一个好的决策者往往对于城市建设的推动更加具有力度，而我作为政府官员，同样也实现了自己的专业梦想。由于各方面的条件和资源更加有效，有很多设想都可以付诸实践，我想这比建筑师在某个项目或区块上发挥的作用更大。我也希望有更多人能对这座城市发挥更多正向的作用。

［采访者：马卫东　文筑国际创始人/上海市建筑学会理事］

P43: SUN Jiwei and Tadao Ando discuss the design for the Shanghai Poly Grand Theatre
P44: The foyer is not only a resting space for audience members; it is also their stage
P46: Autographed letter from Tadao Ando, sent to SUN Jiwei
P47: Photo of SUN Jiwei and Tadao Ando in the Shanghai Poly Grand Theatre

43 页：孙继伟与安藤忠雄见面探讨保利大剧院设计方案
44 页：休息厅不仅是观众入场前的休息场所，也是一个观众演出的空间
46 页：安藤忠雄给孙继伟的亲笔书信
47 页：孙继伟与安藤忠雄合影于上海保利大剧院

# Preliminary Study Process
## 初期探讨

In the initial conceptual phase, the Theatre layout and structure were discussed in relation to the adjacent Commercial & Cultural Centre. The preliminary structural study of the complex was centred on the ovoid shape, as seen in the second sketch. The simple square plane in the third sketch was finally arrived at after combining the results of the preliminary discussion with Poly Property's requirements. The cuboid prototype was then expanded to become a schematic plan for the Theatre. Meanwhile, the idea of arranging the entrance hall and foyer along the main axis was only confirmed after careful consideration.

在初期概念探讨阶段，由于该项目与商业文化中心毗邻，于是从这两者的关系角度出发，就剧院的布局与形体进行了探讨。

在初期形体研究阶段，以初期草图 2 的椭圆形体为中心议题展开了讨论，在综合了业主的要求及探讨的结果，最终生成了以简单的正方形平面为主的初期草图 3，然后再从正方体的原型上裁切出剧院的轮廓，由此创造出公共空间。同时沿主轴线设置入口大厅与休息厅这一基本思想也逐渐被确定。

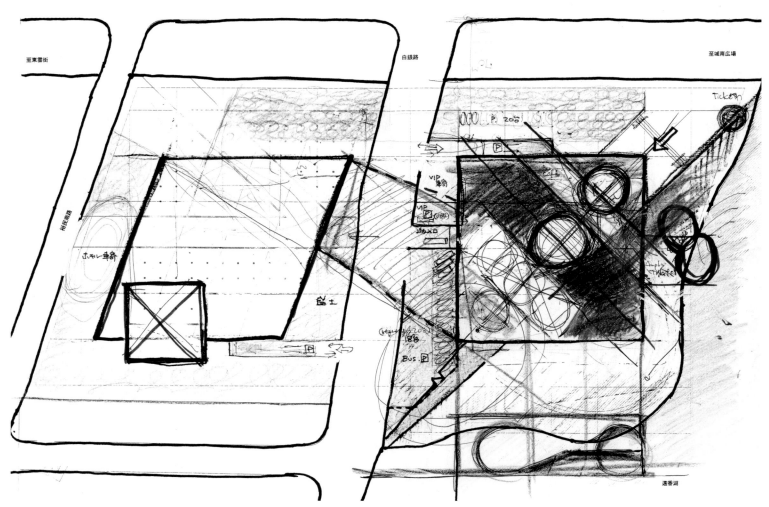

Various options arising from the prototype design in Sketch 3
以初期草图 3 为原型的各种设想

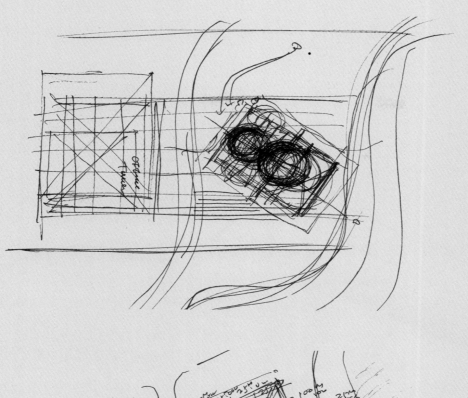

**①**

Sketch 1: The building faces the lake at a 45-degree angle, and is enclosed by an outdoor plaza. The building is nested boxes plane, the main idea was to have the open space surround the opera house complex.

初期草图1：
建筑面朝湖45°设置，其周围设置广场。建筑为套匣式的平面，公共空间在外圈。

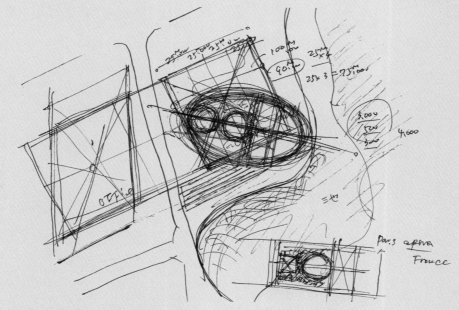

**②**

Sketch 2: An ovoid shape is designed to reflect the curvature of the site. The design features a lake in close proximity to the complex.

初期草图2：
为与基地形状相呼应而设计了椭圆形平面方案，并设置了紧贴建筑物边界的水池。

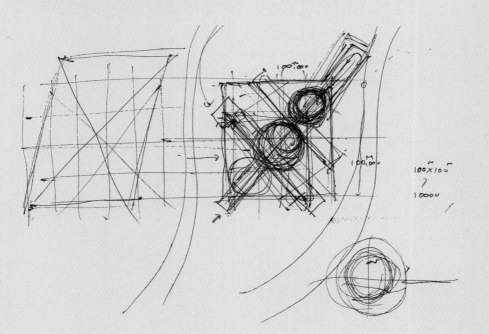

**③**

Sketch 3: Studies of the square design site plan. The main auditorium is placed on the square's diagonal axis. The blank areas in the sketch are set aside for other public buildings and/or spaces.

初期草图3：
正方形平面研究。将主观众厅设置在正方形对角线上，后在空白处插入其他体块。

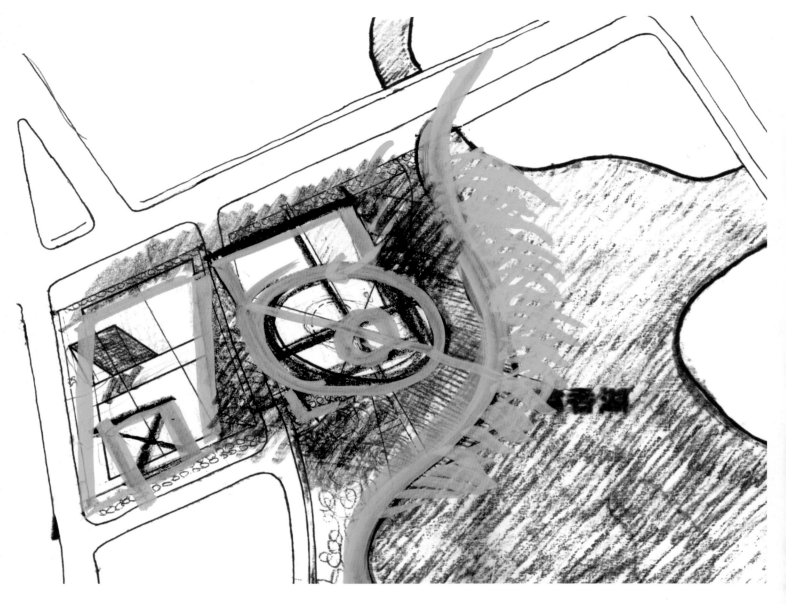

Preliminary study: ovoid plan
初期形体研究阶段的椭圆型平面方案

The original concept was to make a HP-curve under the ovoid plan, using its geometry to emphasize and enliven the ovoid site by virtue of its graceful concavity.
However, due to differences in the elevation of the opera stage and the building complex, finding a functional solution proved to be difficult; therefore, other options had to be explored.

原构想在椭圆平面的下部，根据图纸上的几何关系做出 HP 曲面，该曲面会自然凹出一个微妙的空间，而建筑本身也会因此显得更加灵动。
但由于舞台标高与场地标高的不同，在功能配置方面存在一定的困难，因此只能探讨其他的方案。

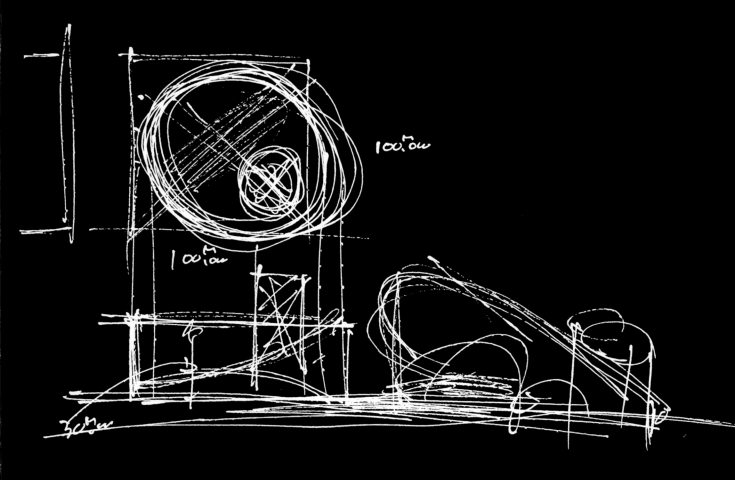

Sketch: Ovoid opera house playfully extends to the lake
剧院的椭圆型草图呈飞跃状态向湖面伸展

# Plans
## 方案评述

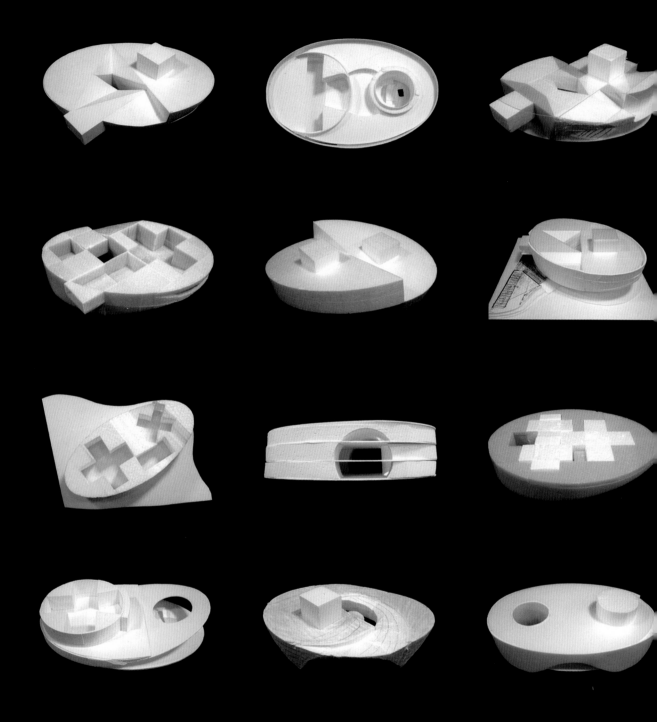

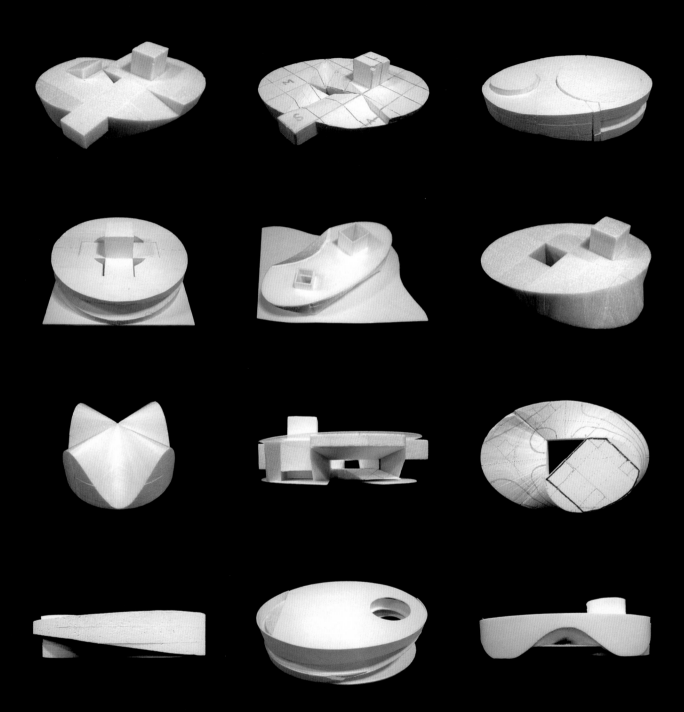

# Plan A
方案A

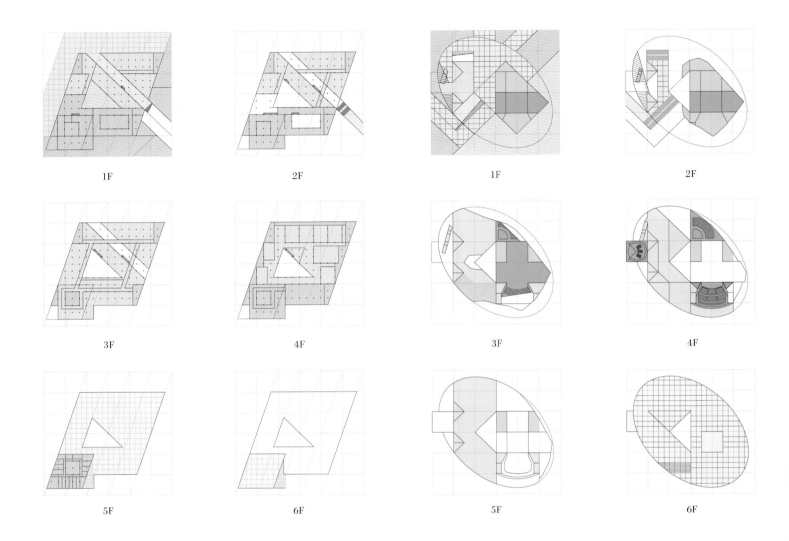

The Theatre is elevated by a series of columns, creating a grand entrance that faces Yuanxiang Lake. The approach from the lake will give people a strong impression. The ovoid volume is able to accommodate various spaces, and the lobby is designed as to afford an unimpeded view of the lake. A major characteristic of Plan A is the lower level below the Theatre, which allows for dynamic and interactive activities.
The form of the upper levels is generated by applying multiple twists to the architecture, creating a dynamic structure whose appearance changes according to the vantage point of the viewer.

剧场由支撑柱向上抬起，形成面向远香湖的"大门"，矗立在湖畔的标志性外观将给人强烈的印象。椭圆形体中能容纳各种功能空间，并且设置了可以将远香湖一览无余的休息大厅。动态的支撑柱空间是该方案重要的特征。

高层的形状是平面的扭曲形态，成为随方向不断变化的都市建筑地标。

# Plan B
方案 B

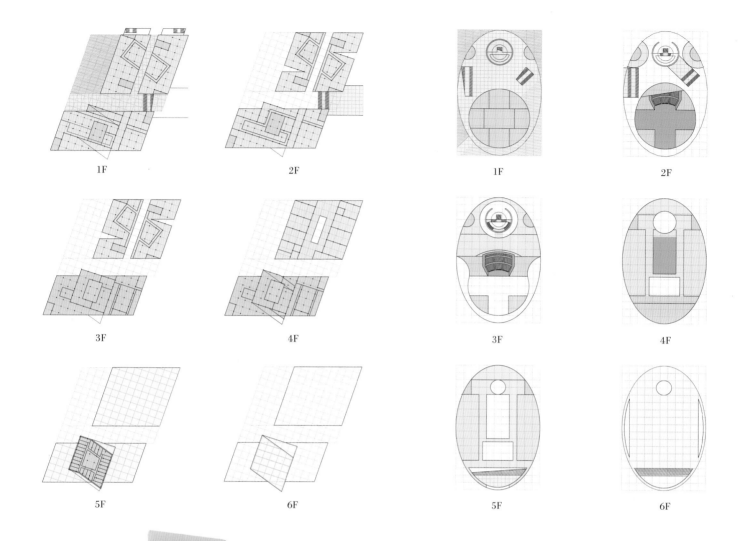

The complex embraces a symmetrical design, with spaces arranged as to reinforce spatial equilibrium. The entrance hall, the foyer, and the main auditorium are aligned along the main vertical axis. The lower level's exterior and ceiling are formed by a curved, three-dimensional surface, whose curvature evokes an embrace welcoming visitors to the complex.
The Commercial & Cultural Centre is divided into multiple tower blocks, with an exterior promenade, which cuts through the central area, linking the spaces to one another. Formal twists are also applied to the centre's upper levels.

剧场的配置是将各个功能空间对称布置；入口大厅、休息厅、主剧场并列在一条轴线上，外墙、支撑柱构成一体性的、动态的三维曲面包容性空间。

文化商业中心采用分栋形式，室外长廊联通中心部分；高层同样采用平面扭曲的形态。

# Plan C
方案 C

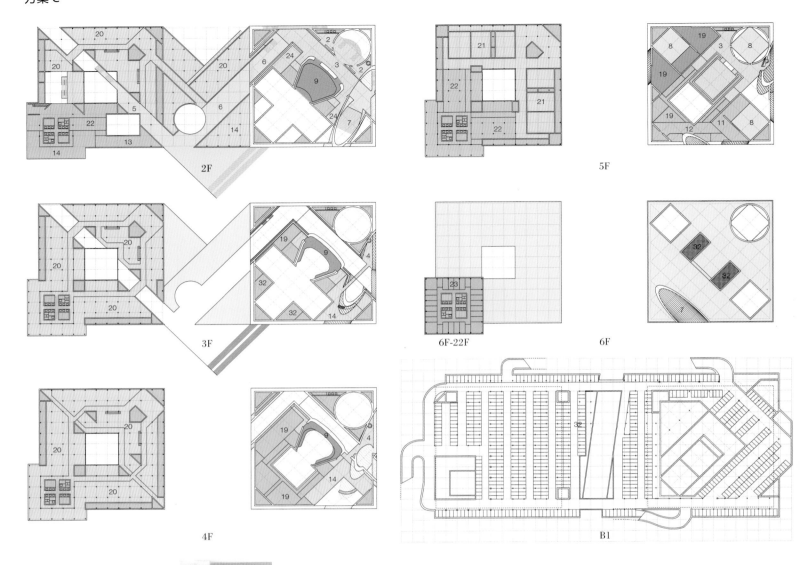

The complex is set on a square plane and accommodates four cylindrical volumes, each with a diameter of 18 metres. A staggered arrangement of the spaces between the cylinders allows for spatial diversity. This effect is accomplished by the specific curvature of the cylinders, as well as by the grid-like design of the plane. This design allows for multipurpose usage of the Theatre complex, and is also feasible in economic and practical terms.

The same concept of having cylindrical volumes cut through the square plane is applied to the design of the Commercial & Cultural Centre. However, the cylindrical volumes have been replaced by square sections, making the structure much more economical. The upper level follows a simple cuboid design, thereby making the square plane the thematic focus of the entire design.

将剧场设置为正方形平面，内部容纳 4 个直径为 18m 的圆筒，圆筒之间通过各种交错形成多样性的空间；通过将圆筒设定成一定曲率的曲面，并且采用网格形的基本结构设计。此外，在完成独特空间的同时，也充分考虑方案的经济性和施工的可能性。

文化商业中心也是采用正方形平面内贯穿圆筒的构成方式，只是这里采用的圆筒改为正方形的断面，使其获得更加经济性的结构体系；高层的形状为单纯的立方体，这是为了将正方形平面贯穿于整个设计方案。

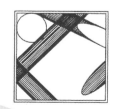

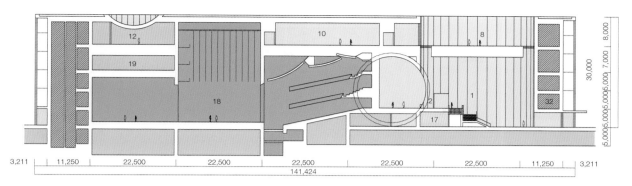

A Section (scale:1/1,000) / 剖面图 A（比例：1/1,000）

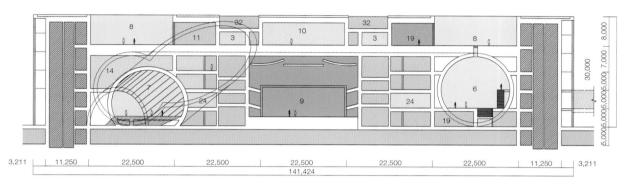

B Section (scale:1/1,000) / 剖面图 B（比例：1/1,000）

Legend:
1. Entrance Hall
2. Lobby
3. Foyer
4. Viewing Gallery
5. Mall
6. Terrace
7. Amphitheatre
8. Court
9. Auditorium
10. Multipurpose Hall
11. VIP Room
12. VIP Lounge
13. Cafe
14. Restaurant
15. Shop
16. Cloakroom
17. Gallery
18. Stage
19. Backstage
20. Commercial & Cultural Centre
21. Cinema
22. Hotel
23. Service Apartment
24. Toilet
25. Parking
26. Bus Parking
27. Loading Dock (Theatre)
28. Loading Dock (Centre of Commerce & Culture)
29. Loading Dock (Hotel)
30. VIP Entrance
31. Lower Entrance
32. Equipment Room
33. Storage

图例：
1. 入口大厅
2. 接待大厅
3. 休息室
4. 观景台
5. 商场
6. 阳台
7. 室外剧场
8. 中庭
9. 观众厅
10. 多功能厅
11. VIP 室
12. VIP 休息室
13. 咖啡吧
14. 餐厅
15. 商店
16. 衣帽间
17. 美术馆
18. 舞台
19. 后台 (后勤空间)
20. 商业文化中心
21. 电影院
22. 酒店
23. 服务式公寓
24. 卫生间
25. 停车场
26. 客车停车位
27. 大剧院货运入口
28. 文化商业中心货运入口
29. 酒店货运入口
30. VIP 专用入口
31. 次入口
32. 机械室
33. 仓库

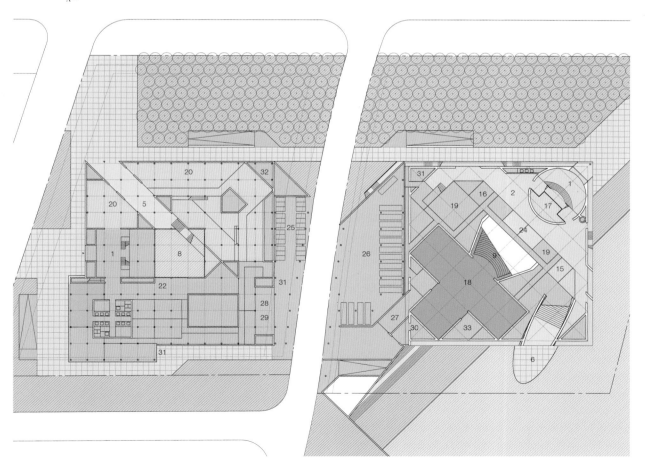

Basic Plan (scale:1/2,000) / 基本平面（比例：1/2,000）

# Cultural Kaleidoscope
文化万花筒

Interview with Tadao Ando
采访安藤忠雄

Interviewer: MA Weidong / 采访者：马卫东

**Q:** As we know, you are good at designing cultural buildings, especially art galleries and museums. There are many well-known projects around the world designed by you. However, this is your first time designing a bigger and more formal theatre like the Shanghai Poly Grand Theatre. What did you think of this kind of cultural building when you decided to accept the project?

**A:** In my opinion, no matter if it is residential housing, a commercial facility, or an art gallery, what I always focus on is creating a space where people can congregate. This is essential for an architect. Even if it was my first time designing a formal theatre like the Shanghai Poly Grand Theatre, in which many technical problems had to be solved, I felt the same impulse to create a moving space, as what I am still exploring is the essence of architecture.

**Q:** Designing a theatre requires advanced technologies and techniques. There are industry standards for acoustics, lines of sight, stage machinery, etc. In fact, it is more important to define the civic nature of a theatre through the various activities and spatial experiences that people encounter when they are not attending a performance, than it is to reach industry standards. The civic nature of a theatre endows it with different characteristics and personalities. What is your view on the civic nature of a theatre?

**A:** In a theatre, we need not only the basic audience space but also a platform where the audience can share its feelings and enjoy an advanced degree of socializing. To ensure that spectators will experience something special during the intermissions, we incorporated unique architectural elements in the entrance hall and other public spaces in the theatre.

**Q:** From your essay *The Challenges of Creating a Cathedral to Culture* you mentioned the world "drama." Is it the same as "civic nature" in the context of the theatre? What are the differences?

**A:** All kinds of people gather in a theatre to partake in experiences that they cannot have in everyday life. Therefore, I proposed the idea that the theatre be open to the public. Not only can visitors share the public space, but also the experiences which can only be enjoyed in this setting. All of these ideas are connected to the shape of the space.

**Q:** Do you recall discussing the topic of civic spaces with Mr. Sun during the Shanghai Poly Grand Theatre lecture on 22 December 2014? You mentioned that even when the theatre wasn't hosting any performances, it would be open to the public, making it the living room of Jiading District, a public space that residents in Jiading would be proud of. Is this what you mean when you talk about the civic nature of a theatre? What responsibilities do you think a theatre, as a cultural space, should have?

**A:** I always strive to design a theatre that is open to the public, even if it is the public property of the municipal government. As the area's landmark, I hope it can become the public space that residents love to visit.

**Q:** "Cultural kaleidoscope" is your design concept. The dialogue between people and music, people and culture, people and light, people and nature, and people and people made the various cultural spaces resemble the reflections of a kaleidoscope. Is this the effect you wanted? Could we say that what you wanted was to present the civic nature of a theatre in your design?

**A:** I hope the theatre is not only a space for people, but also a nexus of culture.

**Q:** To ensure the theatre's civic nature, you used a simple and clever design by inserting four cylinders with a diameter of 18m and one

**Tadao Ando**
安藤忠雄

Architect
1969 Established Tadao Ando architecture & Associates
1995 Recipient of the Pritzker Architecture Prize

建筑家
1969 年创立安藤忠雄建筑研究所
1995 年荣获普利兹克建筑奖

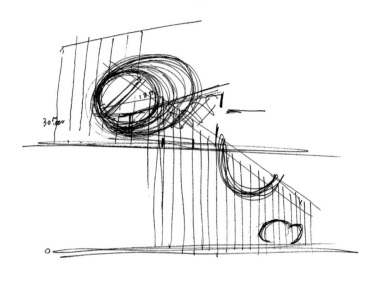

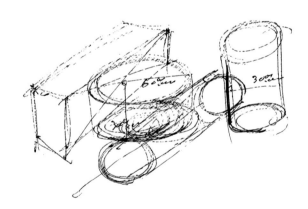

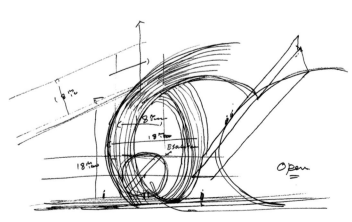

问：我们知道，您特别擅长文化建筑，尤其是美术馆、博物馆，在全世界范围内留下了很多名作。但是大剧院项目，尤其像保利大剧院这种大规模而且正规的大剧院，其实应该还是第一次吧。您当初接受这个项目时，是如何看待大剧院这类文化建筑呢？

答：我一直认为，创造出能让大家凝聚和汇集的场所才是建筑的本质，无论是住宅、商业设施或是美术馆，这都是我一贯的设计主题。像保利大剧院这样正规的大剧院，虽然这次是第一次设计，需要解决的技术问题有许多，但是要做出能让人感动的空间，就这一点而言，和我到目前为止一直在探索的建筑本质没有差别。

问：大剧院设计对技术的要求非常高，如声学、视线、舞台机械等，都有一定的技术标准。其实，评价一个大剧院设计的好坏，除了特定的声学等技术达标外，大剧院的公共性，即人在观演以外的时间里，在大剧院的各种活动及空间体验，更为重要，它是带给大剧院不同个性及特色的载体。您是如何看待大剧院的公共性的？

答：大剧院除了基本的观演功能外，还需要具备让来访的观众能相互共享感动、提供高层次的社交交流的场所这样的功能。在观演以外的时间里，为了能给予观众非日常性的感动，在入口大厅及休息厅等公共空间里，即使采用通常的建筑材料，也希望能向观众提供未曾有过的空间体验。

问：在您的文章《文化殿堂的挑战》中，也提到过"剧场性"一词，这和大剧院的"公共性"是同一个内容吗？他们之间有何不同？

答：在大剧院里，各种各样的人汇集在一起，共享那些在日常生活中无法体验得到的非日常体验。因此，我在设计中，提出了面向市民开放的大剧院这样的主题，让来访的观众除了可以共享公共空间以外，还可以共享到只有在这里才能享受到的空间体验，这些都和空间形状产生关联。

问：有关大剧院的公共性，记得 2014 年 12 月 22 日，您在保利大剧院的讲演会开始之前，和孙继伟先生有过论述。当时您说，那些公共空间在没有演出的时间段，仍然对外开放，对市民开放，使得大剧院承载着嘉定区的公共客厅的功能，让嘉定区的市民有一个值得自豪的公共活动场所。这就是您认为的大剧院的公共性吗？您觉得大剧院这样的文化设施应该承载怎样的社会责任？

答：在设计中，我一直在追寻向市民开放，作为区域的公共财产而存在的大剧院的形态。作为地域的名片，希望她能成为市民们喜爱的公共设施。

问："文化万花筒"是您在设计中提出的设计理念，在公共空间里，人和音乐，人和文化，人和光线，人和自然，人和人之间的交汇融合，如

cylinder with a diameter of 30m at different angles into a tube 100m by 100m. They inform the design of the theatre's various spaces, such as the entrance hall, the lobby, spaces that are half-indoors and half-outdoors, the semi-outdoor amphitheatre, the cafeteria, etc., all of which are very enchanting. Why did you choose to employ the principle of subtraction in your design strategy at the start of the project?

**A:** I chose to enclose the space using the representative architectural materials of the 20th century: concrete and glass. The position of the simple cylindrical spaces were set to correspond to the city's axis and the flow of people. The intersections of the cylinders are the spaces in which people can be in communion with nature. I tried my best to use simple, abstract shapes to present the holy and stately spatial effect.

**Q:** Your characteristic use of fair-faced concrete in your architecture has made you very famous, especially since it requires you to adopt construction processes which are not only more difficult, but require great precision. It is said that in many of your projects outside of Japan, you had deliberately avoided the use of fair-faced concrete in their designs. However, you used quite a large amount of fair-faced concrete in the Shanghai Poly Grand Theatre project. At the beginning, did you not worry about the quality of construction in China? Are you satisfied with the concrete's effect after the completion of the project?

**A:** China is constantly improving its construction technologies. Fair-faced concrete is a material which is both rough and delicate. In this project, the construction worked perfectly. You can feel the warmth of the hands that worked on the fair-faced concrete. The concrete has proven to be quite effective.

**Q:** In fact, the most difficult parts of the construction process were the finished surfaces of the cylindrical spaces. In fact, the surfaces' positioning and segmentation at the intersection of two cylindrical spaces made it quite difficult to be precise, due to the spaces' different axial alignments. How would you evaluate the construction of the Shanghai Poly Grand Theatre?

**A:** I believe the project's degree of difficulty was quite high, but it also a very complete project. I was surprised by the impressive technical expertise of the Chinese construction team. A building cannot be completed by the architect alone. It derives its value from the collaborative efforts of the client, the construction team, and the workers. During the construction process there were many technical difficulties. However, the builders and designers thought, "we want to create a great project that will be the inspiration of future projects." That is why the project is of such high quality.

**Q:** Since your exhibition at the Shanghai Art Museum in 2005, you have had many important projects in Shanghai, Beijing, Guangzhou, Shenzhen, and Hangzhou. Over these past ten years in China, you have witnessed the transformation of Chinese architecture. Could you talk about the transformation of Chinese architecture in relation to the Shanghai Poly Grand Theatre project?

**A:** Over the last ten years, Chinese architecture has developed rapidly in its economic and technological aspects. The Shanghai Poly Grand Theatre is evidence of this rapid development.

**Q:** Last question. The client required a unique, world-famous theatre that would not only stand as cultural landmark in Shanghai, but in China as well. Do you think you have reached that goal?

**A:** My main concept behind the project was to create a cultural gathering space, one that would inspire the pride of the Chinese citizens. Therefore, I cannot say for certain if it has satisfied the client's hopes and requirements. However, the construction team met the precision requirements set out by the architects, and the even surpassed the latter's expectations. From now on, whether the space will become the pride of the residents in the area will depend on the people themselves will embrace their role as citizens and spectators. As the region's architectural landmark, I expect the Shanghai Poly Grand Theatre to become the cultural centre that everyone loves.

(Interviewer: **MA Weidong**  Founder of CA-GROUP /Member of Architecture Society of Shanghai China)

万花筒般折射出各类文化的景象，这是您希望达到的效果吗？是否可以认为在您的这次设计中，其实最想表达的还是大剧院的公共性？

**答**：我希望大剧院不仅仅是接纳市民及外来访者的空间，更是传播文化信息的基地。

**问**：实现大剧院的公共性，您的设计手段非常简单和巧妙，用4个直径18m的圆筒和一个直径30m圆筒，在100m见方的长方体中，以不同的方向、角度、位置形成不同的公共空间，如入口大厅，等候厅，半室内半室外的体验空间，半室外剧场，咖啡厅等，非常有力度和魅力。当初如何会采用这种类似减法式的设计手法？

**答**：用代表20世纪的建筑材料混凝土及玻璃来围合大剧院的演艺空间，对应城市轴线和人流动线的位置插入单纯的圆筒空间。圆筒和圆筒之间交汇形成的空间形成了人们停留、交流、并与自然交融的场所。尽量采用抽象和纯粹的立体造型来演绎神圣、庄严的空间效果。

**问**：您的清水混凝土建筑特别出名，但是施工要求高、精度大。据说您在海外的很多项目，都是因为考虑到清水混凝土施工的难度，在设计阶段就刻意避免使用。在保利大剧院项目里，采用了大面积的清水混凝土，您对中国的施工品质，从一开始就没有担心过吗？实际建成后，您对清水混凝土的效果是否满意？

**答**：中国的施工建造技术真是日新月异。清水混凝土是看似粗放其实颇具细腻表现力的材料，在保利大剧院项目里，施工团队的工作非常完美，浇筑成的清水混凝土既均质又有入手的温暖感，效果非常好。

**问**：实际上在保利大剧院中最难施工的，还是几个圆筒空间完成面的施工，由两个不同方向的筒状空间交汇成的3D空间，无论是定位还是分割，如何确保施工达到一定的精度，都极具挑战，很想听听您对保利大剧院中施工的评价？

**答**：我觉得施工的难度非常大，但完成度非常高。我很惊讶于中国施工人员高超的技术水准。建筑仅靠建筑师一个人的力量是无法完成的，是需要通过业主和施工团队以及工人们的共同努力，才有可能产生出有价值的建筑。这次保利大剧院项目建造过程中的技术难题堆积如山，但是正是因为建设者们有着"为了下一个项目能做得更好，想要做最好的作品"这样的信念，才诞生出如此高品质的作品来。

**问**：从2005年您在上海美术馆的展览会以来，您在上海、北京、广州、深圳、杭州等地都有很多很重要的项目。您在中国的十年，其实也见证了中国建筑近十年来的变化发展。能否请您通过保利大剧院的项目，来谈谈这十年中国建筑有那些变化？

**答**：中国这十年，除了经济外，在技术方面也达到飞速的发展，通过这次保利大剧院的项目也再次印证了这一点。

**问**：最后一个问题，当初业主委托您设计时，希望能够打造一座全世界独一无二的大剧院，成为上海乃至中国的文化地标，您觉得最终实现这个目标了吗？

**答**：做一个能够让市民们值得骄傲的文化信息传播基地，我一直是以此为目标展开设计的。结果来看，是否满足了当初业主提出的希望和要求，我不敢说，但是建设团队很好地完成了设计师当初提出的高要求的施工精度，甚至远远超出设计师当初的预想。今后，保利大剧院是否会如当初设定的目标那样，作为区域的一个核心，成为世界上令人骄傲的大剧院，这还需要同为观众的市民们的共同努力。作为区域的名片，我很希望保利大剧院能够成为大家喜爱的文化设施。

[采访者：**马卫东**　文筑国际创始人/上海市建筑学会理事]

P58: Tadao Ando was being interviewed　　58页：安藤忠雄接受采访
P59: The sketch from Tadao Ando　　59页：安藤忠雄手绘草图
P60: Overlook at the entrance hall of Poly Grand Theatre　　60页：俯瞰保利大剧院入口大厅
P61: The big opening crisscrossed by the cylinder and the facade　　61页：圆筒与外立面穿插相交后形成的大开口

# Design Concept
设计理念

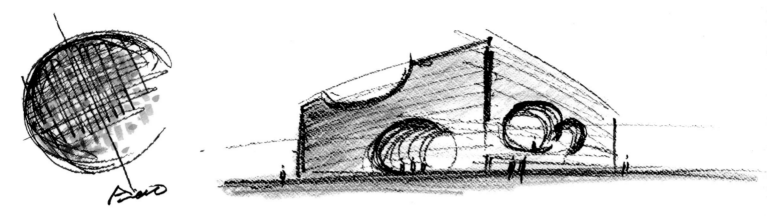

Sketch from Takao Andao: The simple cylinder, utilizing the collisions of nature, culture, and people,
can produce a gorgeous kaleidoscope effect

安藤忠雄的草图——简单的圆筒空间因为人与自然、文化的碰撞，
可以期待犹如万花筒般绚丽夺目的效果

Like kaleidoscope that diffuses light rays to produce a series of dazzling light effects, the opera house complex shall also be cast as a site where nature interacts with people and culture.

This space will become a stage of great excitement and expectations, a place of movement, where people can congregate and engage in interpersonal and cultural dialogue. Behold, the creation of a cultural kaleidoscope.

如同将周围的光线导入，通过漫反射展现绚烂夺目光影效果的万花筒一般，剧场和文化设施也应该作为自然与人、与文化碰撞的华丽场所来对其定位。

市民们将怀着期待、兴奋的心情聚集在这样一个特别的空间场所，将其演绎成一个崭新的充满跃动感的空间。在这里相遇而产生的人与人的对话、人与艺术的对话，构成了剧场空间中各种各样精彩的演出。于是，"文化万花筒"一词随即而生。

# Formal Concept
形体概念

The concept, aside from achieving a grand space of unprecedented proportions, is also an accomplishment in practical feasibility. Here, the following aspects are considered: whether or not the cylindrical design can achieve a simple yet powerful spatial design, in addition to demonstrating a multipurpose functionality.
These cylinders inform the structure's myriad public spaces, such as the entrance hall, lobby, and passageways. The cylinder can be conceptualized as circular forms that move along a straight axis, or as an amalgamation of identical circles. And if it is sliced horizontally, it becomes a straight line. The convergence and divergence of cylinders create complex curvatures, allowing for the creation of fantastic spaces, while maintaining the concept's structural and economic feasibility.

方案设计时，在实现前所未有的壮观空间的同时，也仔细分析了现实中形体规划的施工可能性。在这里，需要考虑管状空间组合的效果是否能形成单纯且具有力量感的空间，以及如何营造出充满多样性的建筑空间。

一个个简单的圆筒，它们构成了入口大厅、休息厅、通道等公共的功能空间。圆筒是圆形沿着直线移动的轨迹，可以考虑为相同直径的圆的集合，如果水平放置切片，它就是直线。圆筒与圆筒的咬合相交生成了复杂的曲线，通过这样的设计，在兼顾结构的合理性和经济性的基础上，创造出意想不到的、各种奇特的空间。

 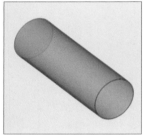  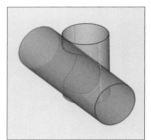 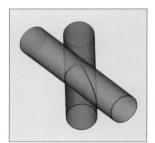

The effect of light passing through openings in the intersections between the cylinders and the main structural complex, as well as the intersections between the cylinders themselves
圆筒与建筑外框相交以及圆筒与圆筒相交而成的曲线及开口的效果

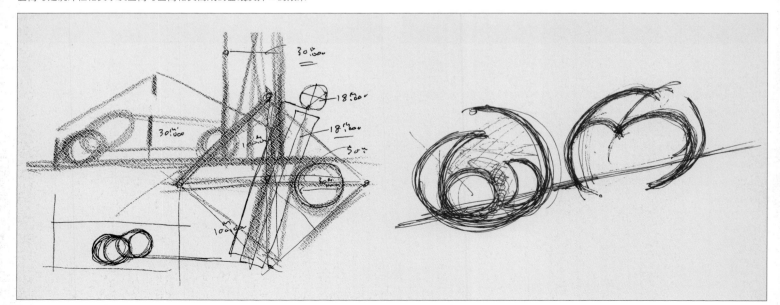

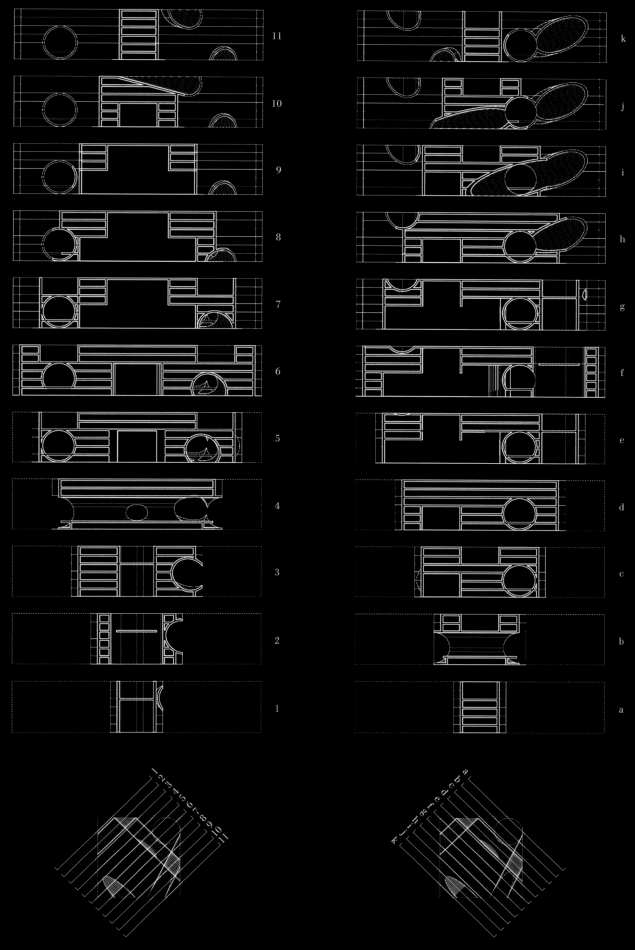

Cross-section views of the theatre
大剧院的横向切片

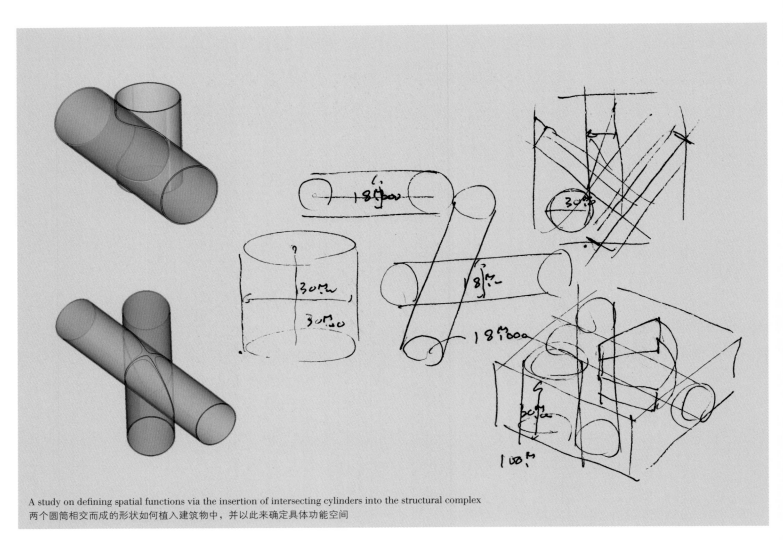

A study on defining spatial functions via the insertion of intersecting cylinders into the structural complex
两个圆筒相交而成的形状如何植入建筑物中，并以此来确定具体功能空间

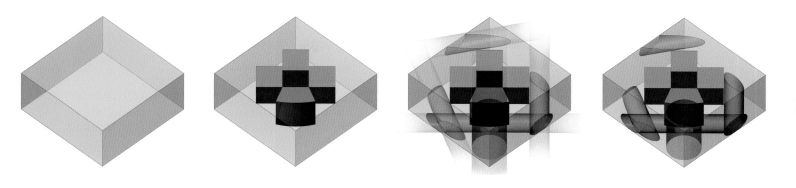

The main auditorium is enclosed within the cube;
cylinders are inserted into the surrounding areas to create a diverse array of open spaces
在长方体体块中设置主观众厅，并在其周围以各种角度插入圆筒，裁切出各种挑空空间

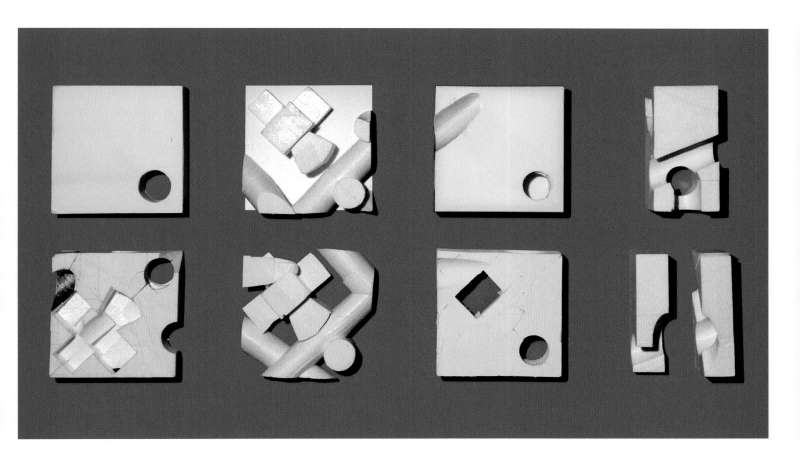

By having the cylinders dissect the cubical structure of the theatre complex, a vast array of interior spaces can be created. Similarly, corresponding oval openings can be generated by placing the cylinders at varying angles with the exterior wall. These oval openings can be conceptualized as the cylindrical space's projections. To facilitate the entry of natural light, large openings are fashioned into the complex's four façades.

将圆筒贯穿于形体，由此裁切出内部空间，而圆筒与外墙的咬合，形成了与各个轴角度相对应的椭圆开口。这些开口形状，也可称作圆筒空间的投影，并在外部四个立面上设置可引入大自然的大开口装置。

# Spatial Composition
空间构成

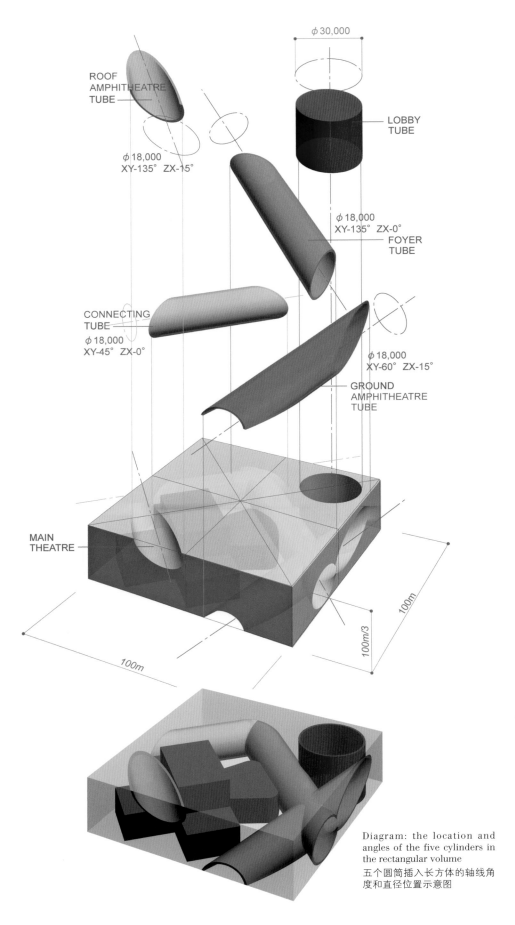

The basic building blocks of the structure consists of rectangular volumes that are 100m long, 100m wide, and 34m high. The main auditorium is positioned along the diagonal axis of the square plan, near one of the corners; hence, most of the open space will be found at the opposite corner, which is where the auditorium entrance is situated. The structure's various spaces are designed by having the cylinders (18m in diameter) cross-section the structural space. This technique then informs the creation of the entrance hall, foyer, outdoor terrace, the open-air theatre, and the overpass connecting the theatre complex to the commercial tower. On a larger scale, the crisscrossing of these cylindrical forms constitutes the overall structure. The apertures and curvatures arising from the cylinder-cylinder and cylinder-façade intersections are the results of countless experimentation and adjustment. The figures in the diagrams represent the axial angles and diameters of each cylinder, with the final angle of the entrance cylinder adjusted to 30m. While the cylinder itself is a relatively simple geometric form, the intersections and connections that arise from the juxtaposition of these forms allow for rich spatial variations. This richness, in turn, transforms this theatre complex into a public stage, where people can engage in all sorts of cultural activities.

建筑的基本形体是长宽各约100m、高34m的长方体。沿着正方形底面对角线并靠近其中一个角设置主剧场，因此大部分的空白空间就会出现在同一轴线的对角处，即观众厅的入口部分。采用直径18m的圆筒进行交错切割的挑空手法，将该空间自然分割，用以规划不同功能的公共空间。从五个圆筒中划分出入口大厅、休息厅、室外平台、地面和屋顶的露天剧院以及通向商业楼的天桥，并通过这些立体交错的空间构成整座剧院。此外圆筒与圆筒交错形成的曲线和开口，还有圆筒与外立面相交后切出的开口效果都是经过不断验证、不断调整后摸索出的平面配置。分析图中的数字分别表示各个圆筒的轴线角度和直径，入口大厅圆筒的直径最终调整为30m。虽然圆筒本身只是一种轴线清晰的简单几何体，但通过圆筒的立体交错、相互连接后，创造出了意想不到的空间演绎。由此形成的公共空间也成为一个舞台，能够引发人们积极地开展各种文化活动。

Diagram: the location and angles of the five cylinders in the rectangular volume
五个圆筒插入长方体的轴线角度和直径位置示意图

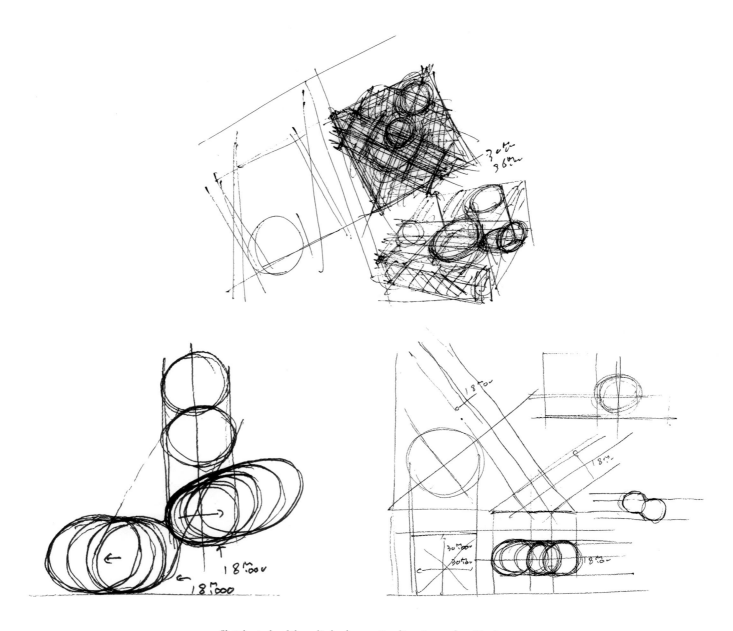

Sketch: study of the cylinders' respective diameters and positioning
圆筒位置及直径的讨论手绘稿

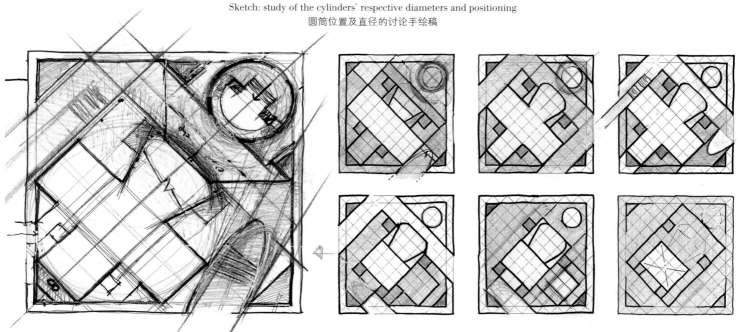

Size of the outer rectangle: 100m × 100m × 34m (*H*); cylinder diameter: 18m
长方体外形参数是 100m × 100m × 34m(*H*)，圆筒直径是 18m

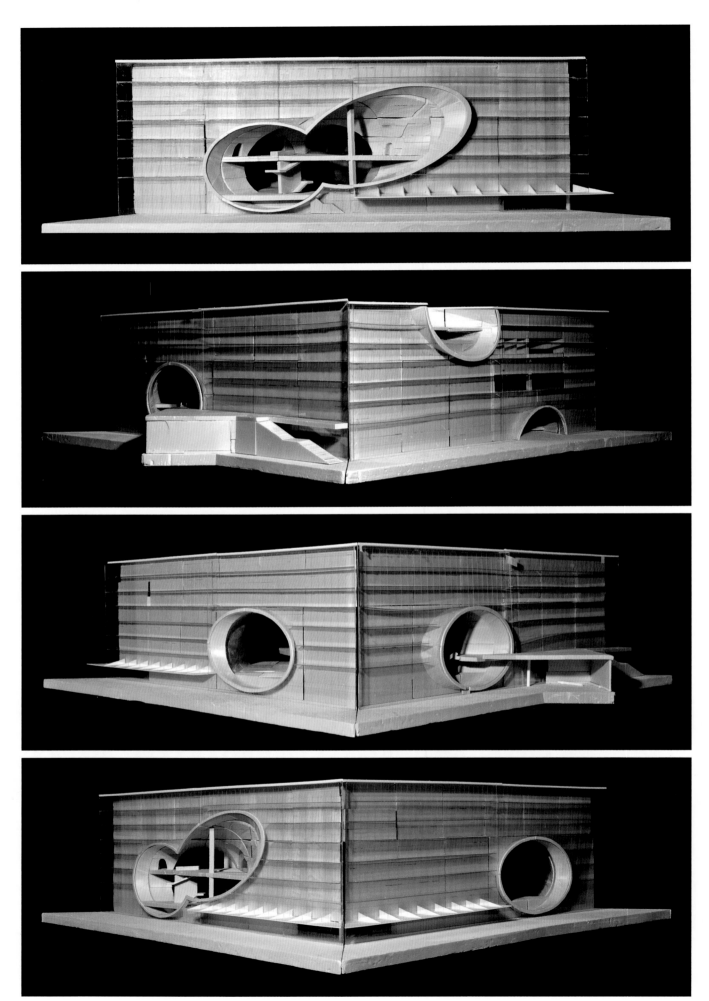
Model of cylindrical forms crisscrossing the façade at varying angles / 不同角度的圆筒与立面交汇的模型

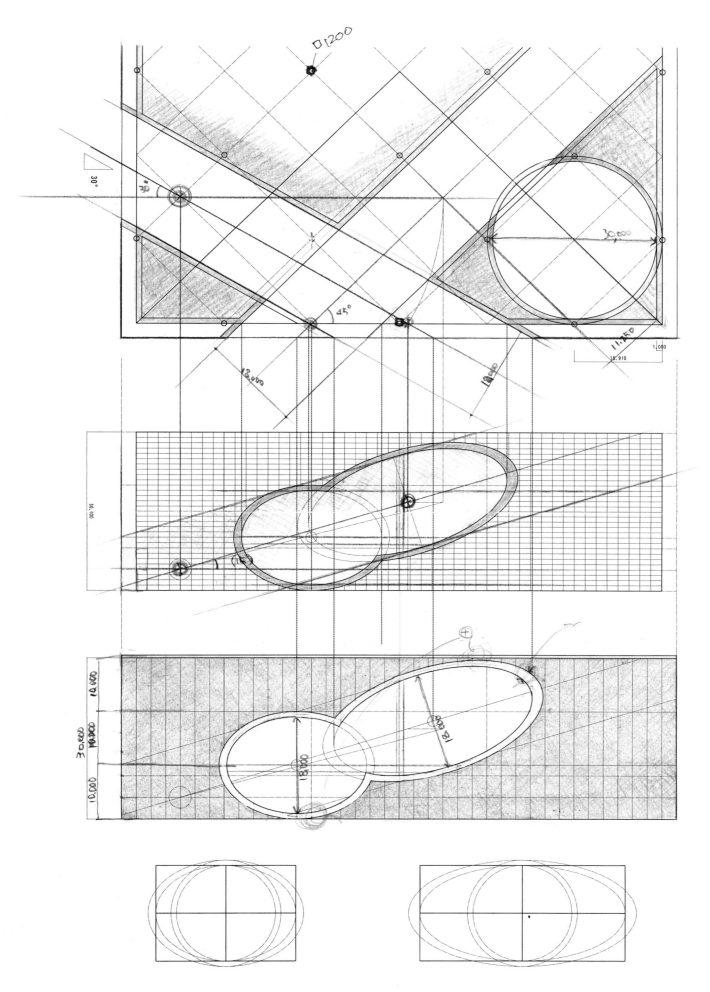

Studies of the position and shape of the apertures arising from the intersection of cylindrical forms with the façade
对圆筒与外立面相交而成的开口位置和形状的研究

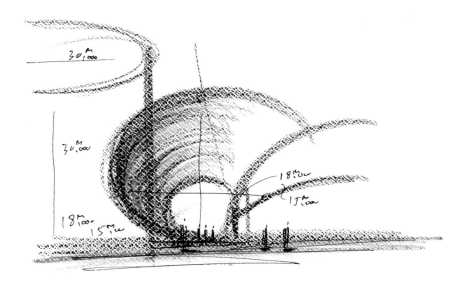

In order to emphasize the cylindrical spaces as the stage for visitors, the cylindrical spaces of the main entrance and the foyer are set at a perpendicular angle. The foyer then extends horizontally into the main auditorium. The smart connection between the entrance and the auditorium is the one brilliant characteristic of this design, creating fluid movement from one space to another. To maintain the foyer's spatial equilibrium, the diameter of the entrance cylinder is set at 30m. The cylinder comprising the foyer is 18m in diameter and over 90m in length, and traverses the structure from the façade on Baiyin Road to the eastern façade facing the lake. Flanking the foyer are two outdoor terraces, which are also formed by the complex interplay of other cylindrical forms.

将剧院的观众席部分贯穿于休息厅的圆筒空间之中，再使休息厅与入口垂直的圆筒空间相交。这些相互连通的灵动空间是此建筑最具魅力的闪光点，是属于来访者的"舞台"。考虑到与休息厅的平衡性，将入口圆筒空间的直径定为30m。构成休息厅的是一个直径为18m，长度超过90m的圆筒，其从白银路一侧的正立面贯穿到远香湖一侧的东立面。圆筒空间的两端是室外露台，此处又是由另外的圆筒相互贯穿而生成的复杂空间。

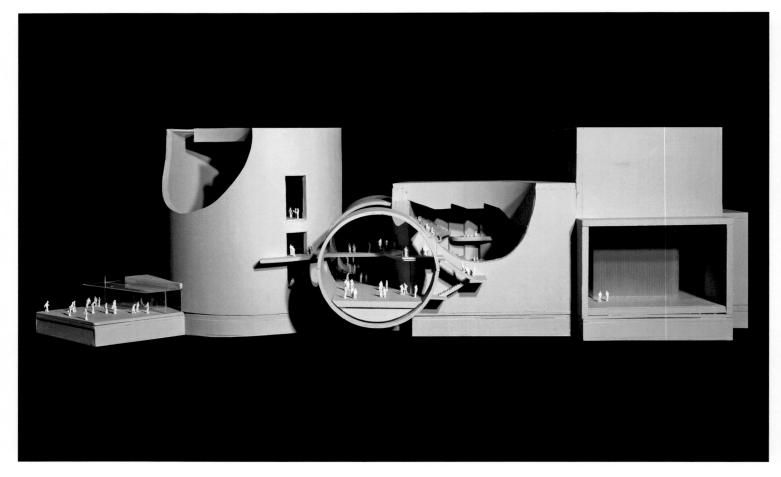

P72: Top, Studies of the cylindrical spaces for the entrance hall and the foyer. The final measurements were set at 30m for the diameter of the entrance hall and 18m for the other cylindrical forms; Bottom, Model showing the spatial continuity linking the entrance hall, the foyer, and the main auditorium
P73: Model studying the spatial relationships between the cylindrical space

72 页：上，圆筒空间的研讨草图，最终确定入口大厅为直径 30m，其余为直径 18m；下，展示了入口大厅、观众厅和舞台连续关系的模型
73 页：研究圆筒空间的工作模型

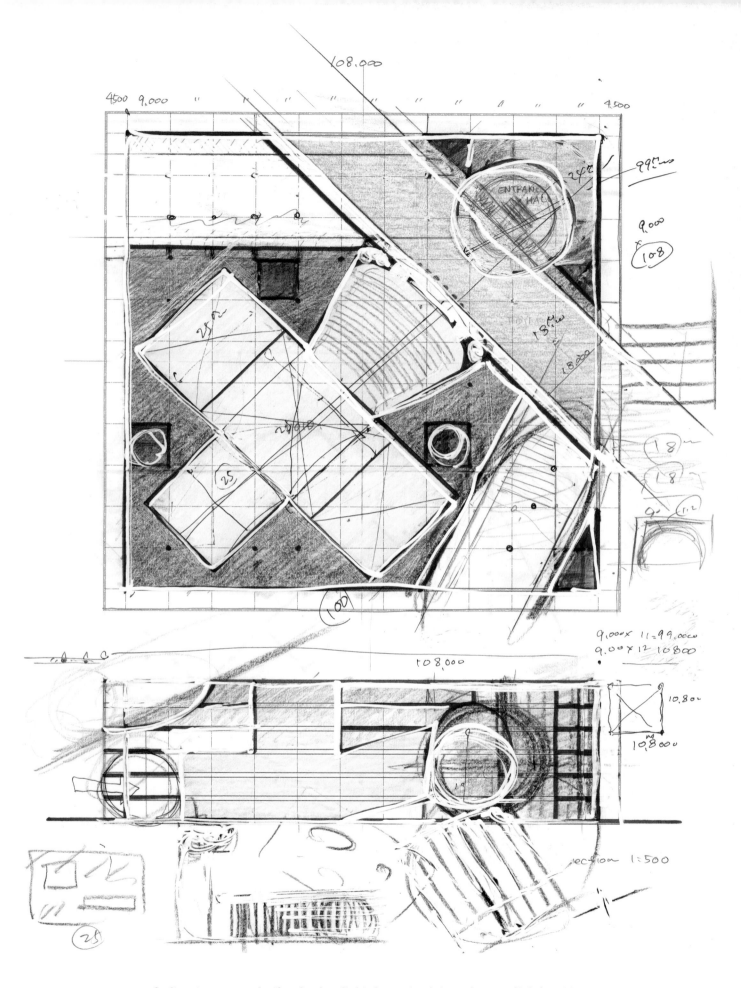

Studies using a structural grid to plan the cylindrical spaces in relation to the core cylinder's position
根据结构柱网中的核心筒位置研究圆筒空间的设置

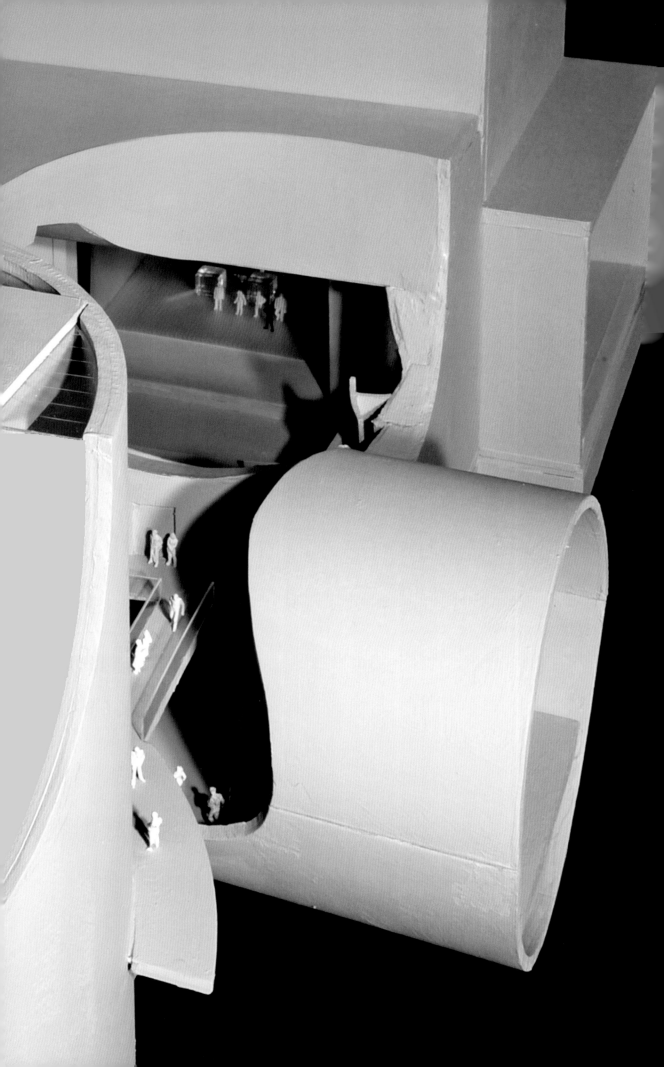

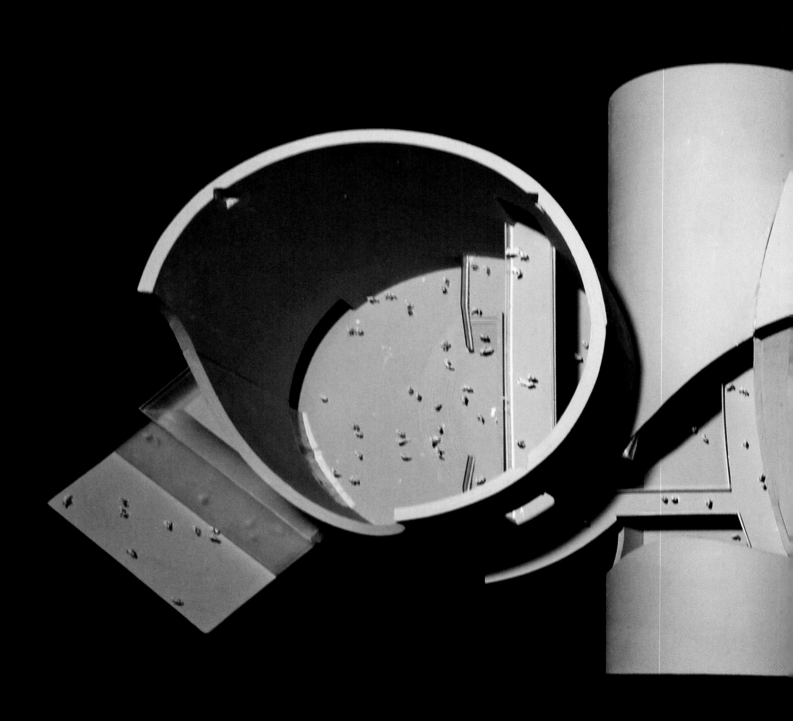

P74: Working model
PP76-79: Working model of the cylindrical spaces
74 页：工作模型
76-79 页：研究圆筒空间的工作模型

# Design Concept
设计方案

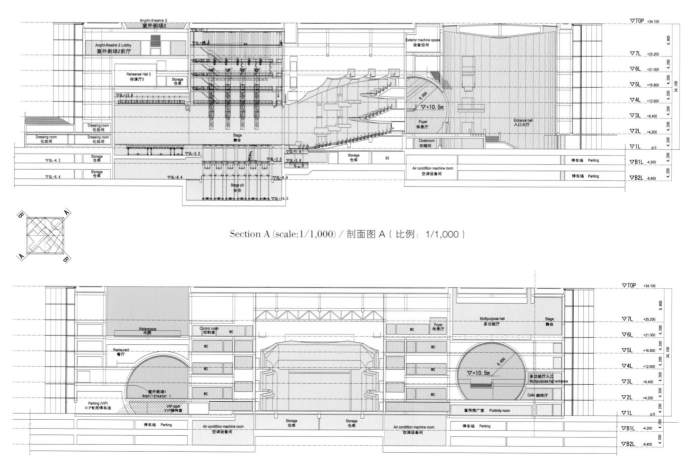

Section A (scale:1/1,000) / 剖面图 A（比例：1/1,000）

Section B (scale:1/1,000) / 剖面图 B（比例：1/1,000）

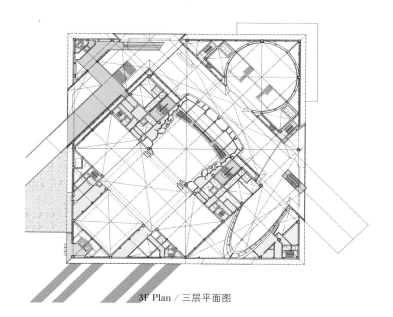
3F Plan／三层平面图

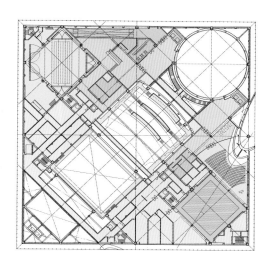
6F Plan／六层平面图

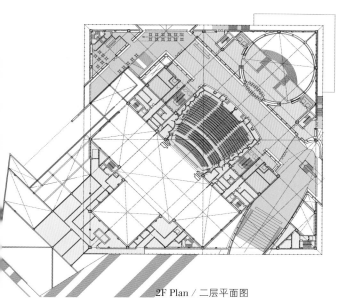
2F Plan／二层平面图

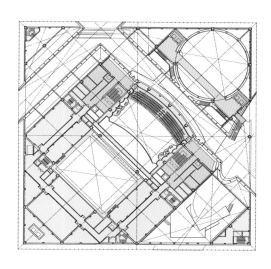
5F Plan／五层平面图

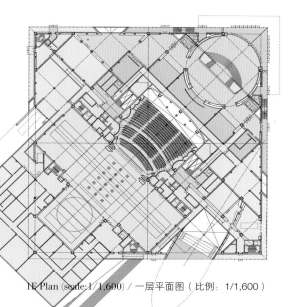
1F Plan (scale: 1/1,600)／一层平面图（比例：1/1,600）

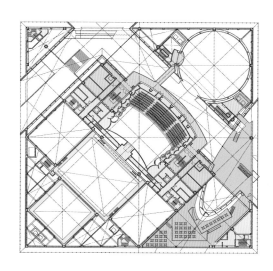
4F Plan／四层平面图

Entrance Hall
入口大厅

The entrance hall, measuring 30m in diameter and 30m in height, is an enormous space. The material forming the cylindrical wall was initially planned to run along a vertical axis, but was later changed to a louver design running along a horizontal axis to match the structure of the foyer. Half of the entrance area is occupied by the stairway leading up to the foyer, imbuing the space with various stage effects.

入口大厅是直径为 30m、高度为 30m 的巨大空间。构成圆筒状墙壁的线形材料起初决定为垂直方向，但为了使其与休息厅以相同构件和相同手法收口，在圆周上采用了横向百叶材料。通往休息厅的阶梯占入口面积的一半，形成一个具有舞台效果的空间场景。

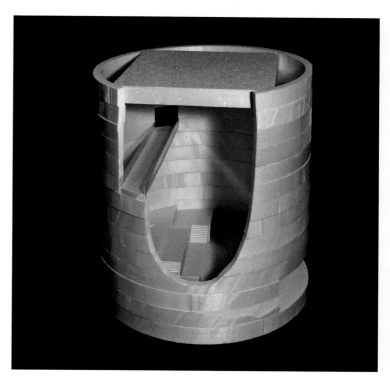

P82: The study model of the entrance hall
P83: Top, The study model; Bottom, The diagram of the entrance hall space discussion
PP84-85: The entrance hall study model
PP86-87: The sketch of the entrance hall

82 页：入口大厅工作模型
83 页：上，工作模型；下，入口大厅空间讨论图
84-85 页：入口大厅工作模型
86-87 页：入口大厅手绘效果图

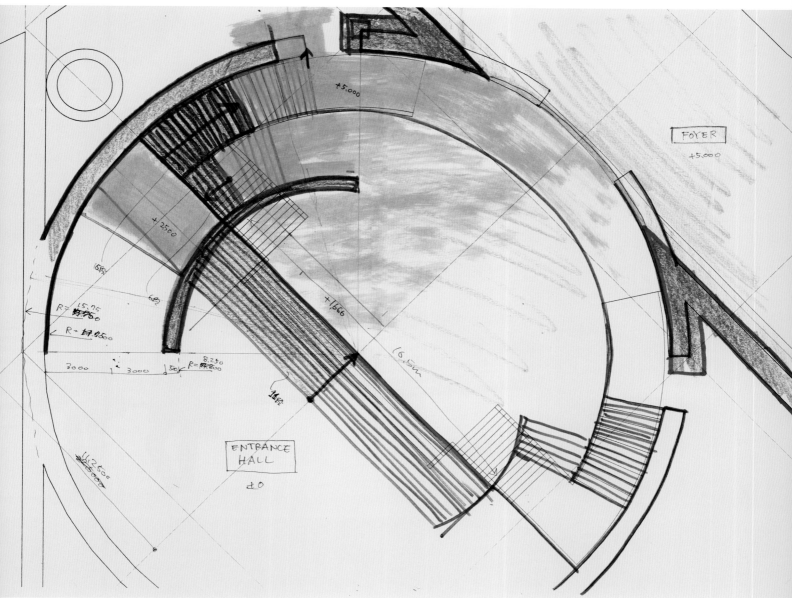

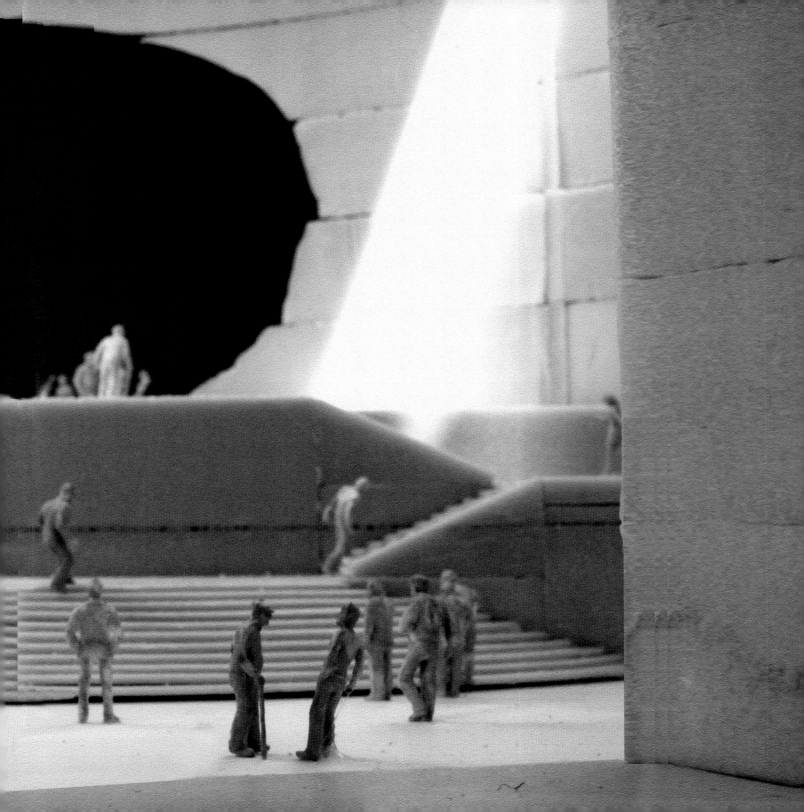

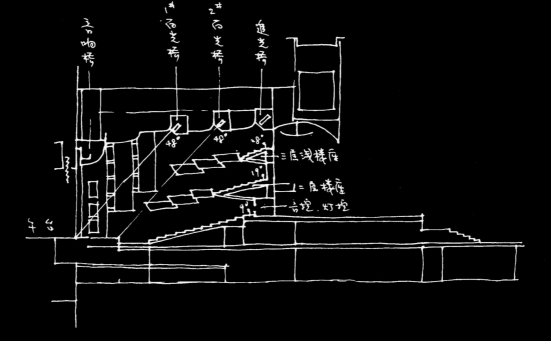

In contrast to the opera houses in Europe, the United States, and Japan, Chinese opera houses tend to place the upper viewing levels/boxes at narrower angles. Initial designs for the auditorium included four-tier seating, but was eventually changed to three levels in the final design.

在中国，注重歌剧演出效果的剧院，与欧美国家和日本相比，通常会将上层观众席和包厢的视角设定得较为偏小。起初探讨过做四层观众席的方案，但最终还是决定做三层。

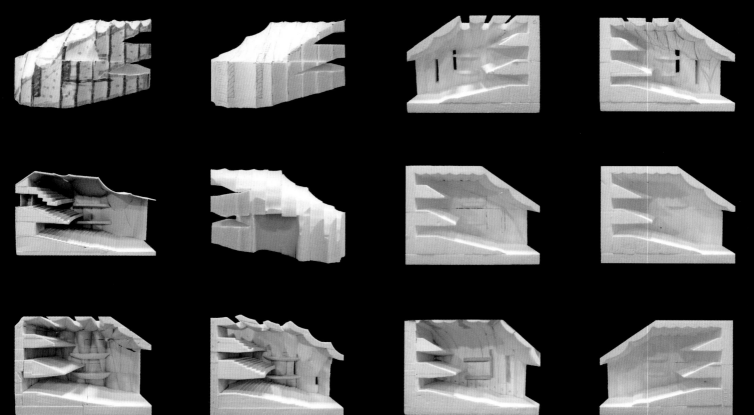

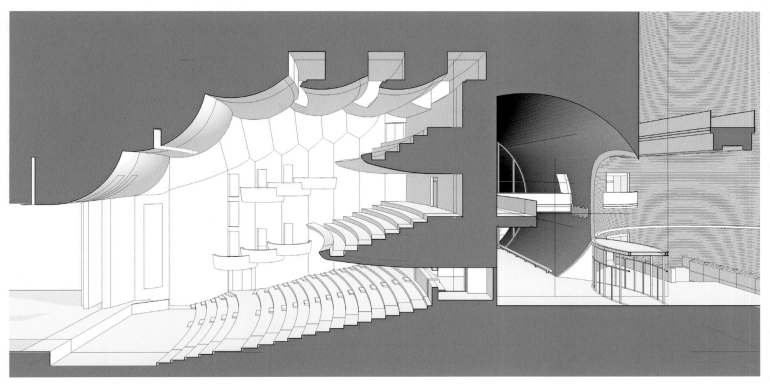
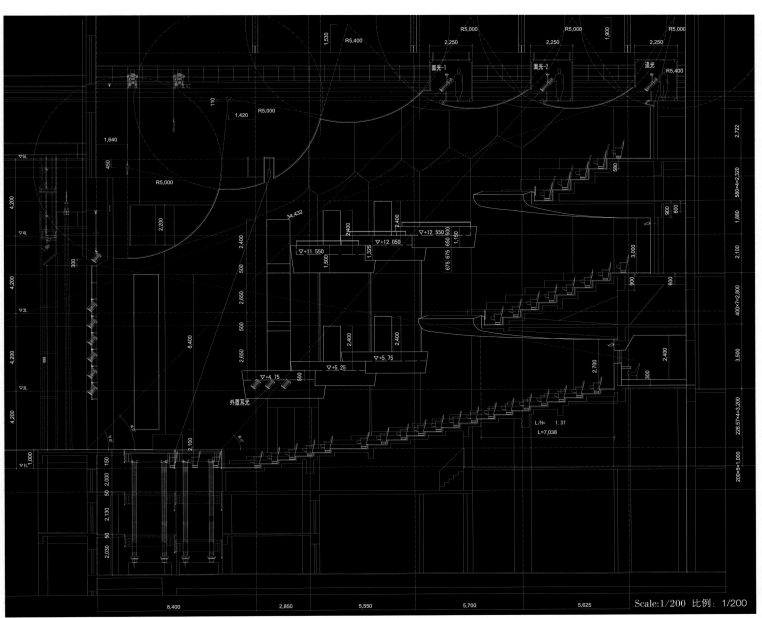

Scale:1/200 比例: 1/200

# Visual Line Analysis
视线分析

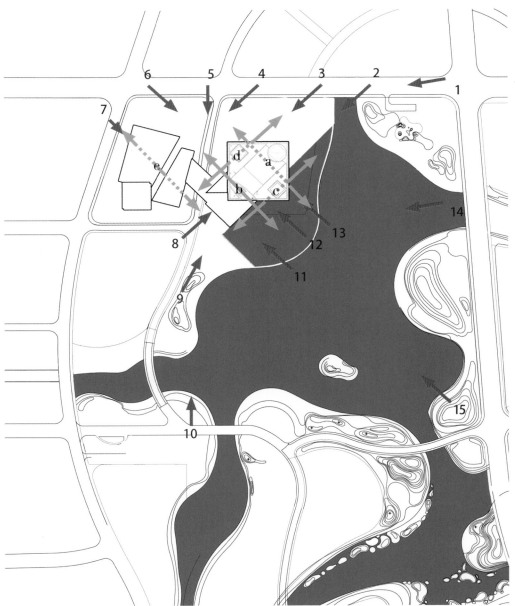

By inserting the cylinders at varying angles into the theatre's rectangular blocks, a diverse range of expressions can be created. In particular, the intersection of the cylindrical form with the façade facing Yuanxiang Lake gives birth to a dynamic exterior. In consideration of the relation between the theatre complex and the Commercial & Cultural Centre, as well as the structures' independent forms, the viewing platform connecting the two structures was set at a 45-degree angle to Yuanxiang Lake. This setup achieves an integrated design that is in harmony with its natural surroundings.

The crisscrossing cylindrical forms generate a dynamic interior, whose stylistic variations evoke rich layers of visual depth and complexity. Views of the outdoor scenery are introduced at various angles, diversifying the range of visual experiences within the complex.

在剧场单纯体块中贯穿的圆筒，通过不同的角度变幻出各种各样的表情。特别在远香湖一侧，圆筒的剖面构成动态的立面，形成标志性的外观。考虑到剧院和商业文化中心的关系，以及它们各自相对独立的形态，将连接两座建筑的景观平台呈 45° 斜角向远香湖展开，与周边的景观进行一体化的设计规划。

交错穿插的圆筒使建筑内部形成了多样的空间形态，让人们在视觉上产生了丰富的层次感和距离感。通过不同的视线角度将周边的风景纳入其中，剧院内看到的风景也会随之出现各种各样的轮廓。

Exterior / 外部视线

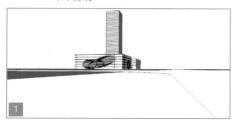
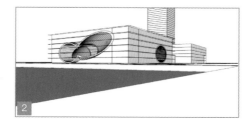
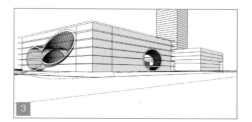
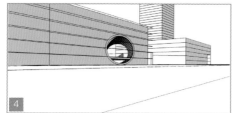
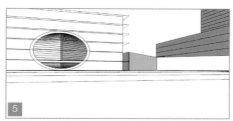
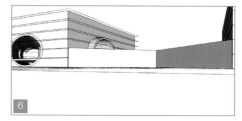
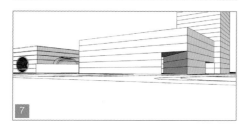
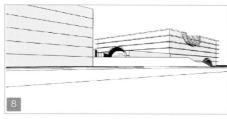
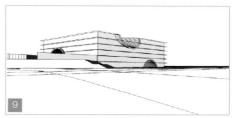
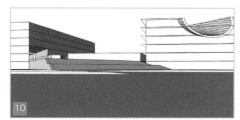
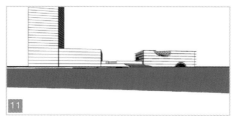
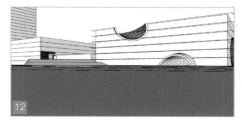
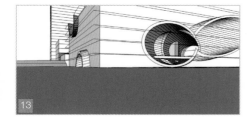
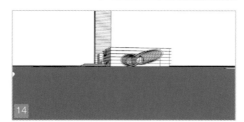
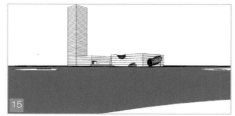

Interior / 内部视线

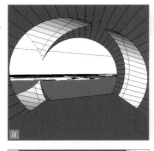
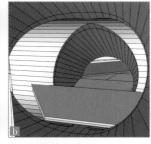
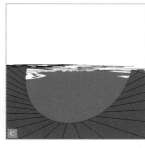

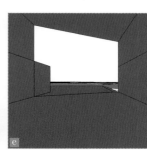
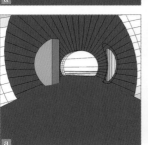

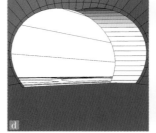
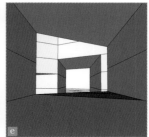

| Foyer tube | Grand amphitheatre tube | Terrace amphitheatre tube | Terrace amphitheatre tube | Commercial centre |
| 休息厅筒状空间 | 室外剧场筒状空间 | 屋顶剧场筒状空间 | 连接回廊筒状空间 | 商业中心 |

# Traffic Streamline, Fire Safety, and Greenery Analysis
交通流线、消防、绿化分析

**Traffic streamline analysis:** The theatre's main entrance is set along Baiyin Road, with the VIP entrance positioned to the east. The main entrance of the Commercial & Cultural Centre is set in the northwest, at the intersection of Baiyin Road and South Yumin Road. The hotel's main entrance is on South Yumin Road in the west, while the entrance to the hotel apartments is to the south.

The theatre's loading bay is positioned at Huanhu Road, while the loading bay for the Commercial & Cultural Centre and the hotel is underground, directly below the ramp in the south.

The viewing platform is designed to facilitate the movement of pedestrians between the theatre complex and the Commercial & Cultural Centre, thereby linking the lively atmosphere of the two spaces. The platform is also connected to the lakeside walkway and plaza.

**Fire safety analysis:** To facilitate the movement of firefighting vehicles, a 4-meter wide passageway and a reasonably large firefighting platform are set around the complex. Under normal circumstances, the space will function as pedestrian walkways.

**Greenery analysis:** A green belt is planted in north, along Baiyin Road, forming a larger green corridor with the surrounding area. Green design has also been incorporated into the complex, including green roof, natural grass slopes, a copse of cherry blossoms, and various shrubs, undergrowth, and aquatic plants. The complex is to serve as a gathering place where people can come into dialogue with nature.

交通流线分析：大剧院的主入口设置在白银路一侧，在大剧院东侧还特别配置了 VIP 专用入口。商业文化中心的主入口设置在白银路和裕民南路相交的西北侧。酒店的主入口在西侧的裕民南路上，而公寓式酒店则是从南侧进入的。

大剧院的后勤服务货运入口设置于环湖路上，而商业文化中心和酒店的货运入口则是通过基地南侧的坡道直接进入地下。

在剧院和商业文化中心的连接处专为行人设置了景观平台，将剧院和商业文化中心的热闹气氛连接了起来。景观平台同时还连接着湖畔的散步道和广场。

消防分析：为确保消防车辆的通行，在建筑物周围设置了宽度为 4m 以上的通道以及合理的消防登高面。而这些通道和空间在平时可以成为市民散步的场所。

绿化分析：基地北侧，沿白银路设置有较宽的绿化带，形成了一条绿色的长廊，将所植灌木与周边道路绿化自然衔接，形成自然形态。

围绕剧院周围展开的绿化设计也丰富多样，有屋顶绿化、自然草坡、樱花林以及各类灌木地被和水生植物，为在这里聚集的人们创造了与自然对话的优美环境。

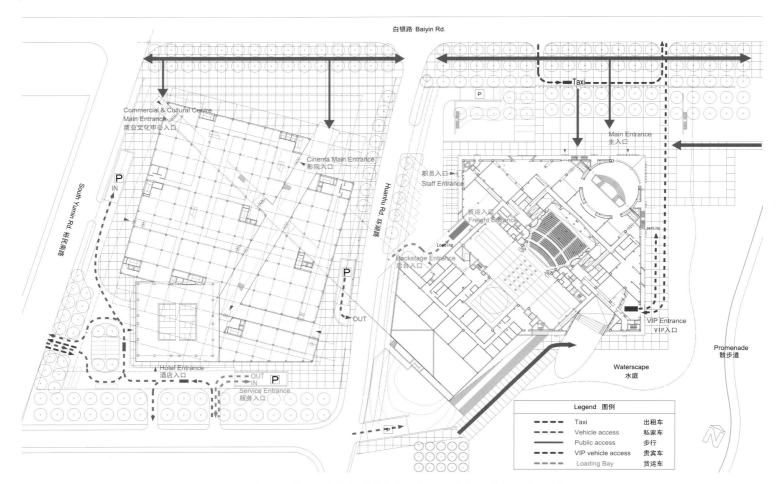

Traffic streamline analysis (scale:1/2,000) / 交通流线分析（比例：1/2,000）

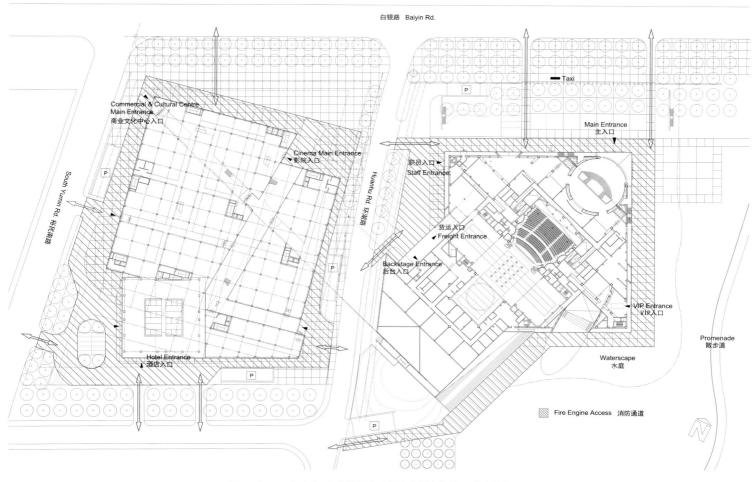

Fire safety analysis (scale:1/2,000) / 消防分析（比例：1/2,000）

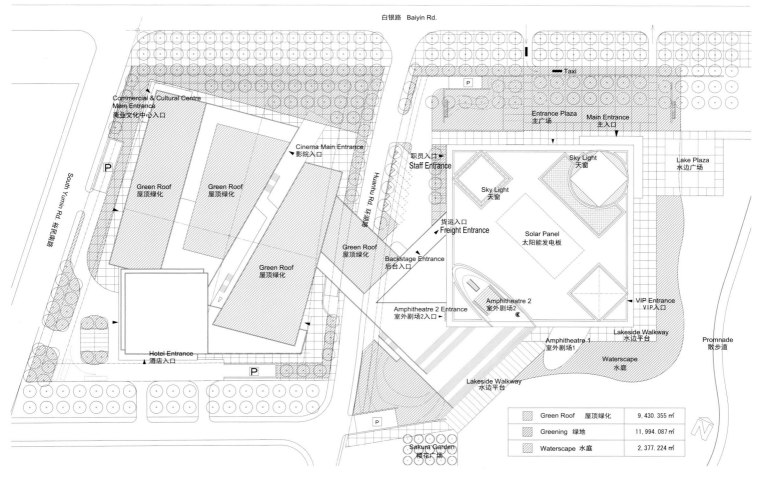

Greenery analysis (scale:1/2,000) / 绿化分析（比例：1/2,000）

# Landscape Analysis
景观分析

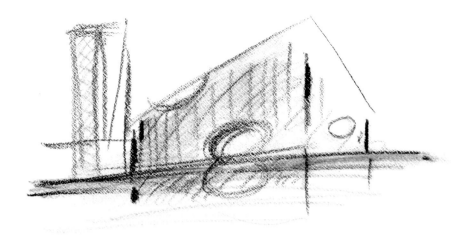

A grid of 5,600mm modules was used during the design of the entrance plaza, allowing for the proper separation of the main entrance passageway and the surrounding trees. The arrangement of plants and the waterscape are designed to distinguish between spaces of movement and spaces of rest.
The waterfront plaza is the sculptural base linking the complex and the arc-shaped bridge. A sense is needed, while maintaining the autonomy of the two spaces. The plaza is defined by its use of grass-embedded granite, which puts the space in harmony with the surrounding scenery. The Waterscape is designed with an emphasis on its relation with Yuanxiang Lake. A visual illusion is generated by connecting the two bodies of water as if they were on the same elevation, when in fact there is a difference of two levels between them.

入口广场设计以 5,600mm 为模数的方格网，适当区分建筑主入口通道与树阵停留空间，利用植物、水景等暗示通行空间与停留空间。

滨水广场是雕塑的"基底"，它在建筑与弧形桥之间，既要联系，又需分隔。设计上用花岗石嵌草的手法界定雕塑空间，又和谐地与外部衔接。

水庭在设计上首先注重与远香湖的联系性，通过 2 级跌水高差设计造成视觉错觉，将地块内的水与湖体"相连"。

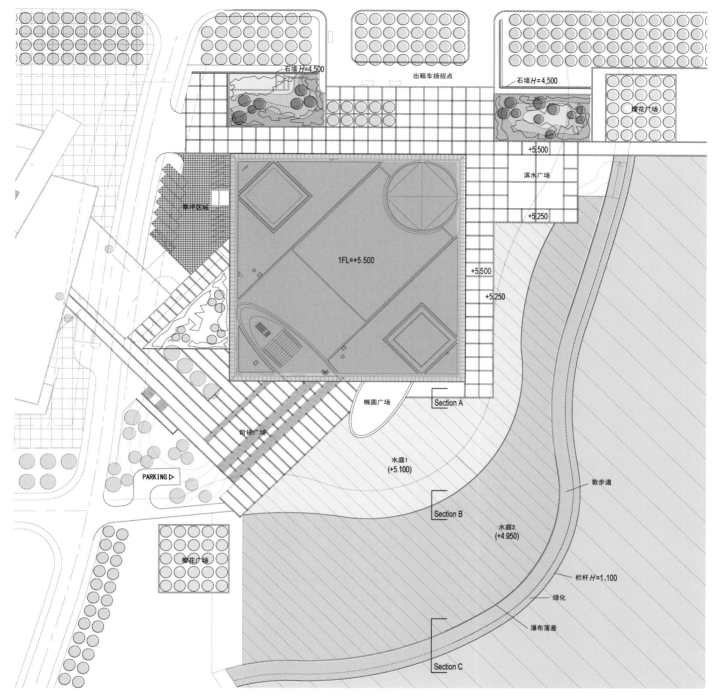

The landscape layout (scale:1/1,600) / 景观布置图（比例：1/1,600）

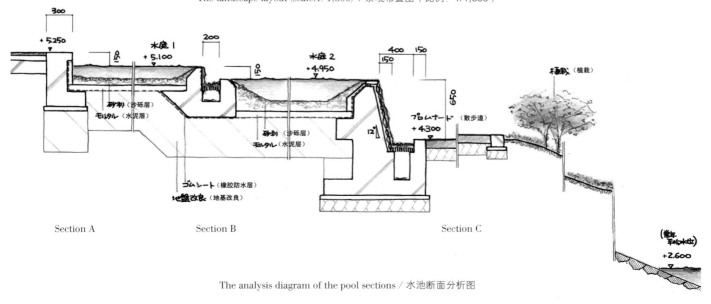

The analysis diagram of the pool sections / 水池断面分析图

# Dynamic Tubes: Ando's Vision Realized in a New Urban Community
跃动的"圆筒"——凝筑于新城中的安藤理念

Kazukiyo Matsuba　松叶一清

## Boundaries Defined by a Square, 100 Metres to a Side [1]

If I were the one to design an opera house in a newly developed suburb of an enormous Chinese city, which was to be the cultural centerpiece and embodiment of the new urban community, how would I have gone about it? When I arrived at the site in question, in Jiading District about one hour by car from the centre of Shanghai, a question and answer exchange on this topic kicked off the largest event to date dedicated to critical appraisal Tadao Ando's work. In designing the Shanghai Poly Grand Theatre, Ando had taken on the heavy responsibility of creating something of immense scale that would aesthetically symbolize the newly developed district, and it was uncharted territory even for an internationally renowned architect who had already designed all manner of cultural facilities in numerous locations around the world.
In terms of building on a vast site, he had already experienced this in designing the Modern Art Museum of Fort Worth in Texas, but this site was already home to an architectural landmark: Louis Kahn's masterpiece, the Kimbell Art Museum [2]. Never one to shrink from a challenge, Ando had shown his characteristic verve in creating a worthy companion to what was recognized as the world's most beautiful concrete building, and this could safely be called the pinnacle of his career thus far. However, there was no existing landmark in the Jiading District to set the stage for this project. Here Ando would be creating new value from square one, and once completed, this project would be a landmark to the district like what the Kimbell Art Museum was to the Fort Worth site: a defining landmark.
One big decision Ando made at the outset was the shape and size of the building footprint: a square, 100m on each side. There was much more space at his disposal on this sprawling site facing on to an artificial lake, and there was nothing inevitable about the selection of a nice, round number like 100m. However, there was a need to demarcate a manageable territory within this vast, ill-defined site, in which to construct his own universe, one he could hold in his grasp. The first step in designing the Shanghai Poly Grand Theatre was to mark the boundaries of Ando's space within the site, and selecting the figure of 100 was a means of establishing this "private universe" as a publicly recognized bastion of universally shared value. To deal with uncertainty or hesitation in the process of creating a structure on a vast site with vague borders, the mathematically-based boundaries of a "square 100m to a side" are a means of anchoring the process steadily to Ando's own chosen scale.
My impression, in any case, was that he approached the project in this way, first of all establishing scale as a defining aspect of this particular Ando creation. When I saw an early sketch executed while he was working out the concept, my conjecture turned to conviction. It was clear that scale was the defining theme Ando tackled during the first stage of design.
In the sketch, there is a cylindrical void in the centre of the side of the building, with another passageway-opening protruding out from its right side so that they formed two linked ellipses. Color-wise, the exterior wall features regularly spaced horizontal blue bands, while the cylindrical tube and the adjoining opening to the right are layered in orange and blue. The blue signifies the water onto which the building faces, and the orange inside the tubes conveys the warm wood-like surface of their interiors. Both along the waterfront and in the interior of the passageway, tiny silhouettes of visitors can be seen.
I was reminded of unbuilt flights of fancy such as Cenotaph for Sir Isaac Newton (1784) by the French visionary architect Étienne-Louis Boullée [3]. Known for creating megalomaniacal architectural designs in the late 18th century around the time of the French Revolution, Boullée placed people as tiny as ants at the feet of his structures in his drawings, to convey their colossal scale. The same ant-people inhabited Ando's sketch. If his groundbreaking Sumiyoshi Row House (1976) was a micro-universe, a condensed crystallization of an individual vision, the Shanghai Poly Grand Theatre was conversely supposed to expand Ando's vision outward to the scale of urban planning, approaching the realm of Boullée-style megalomania. His visionary architectural sketch amply testified to this.
With a square footprint of 100m by 100m, Ando set the height of the building at 34m. This is a height with a certain inevitability not only to Japanese architects, but to Japanese urban dwellers in general. The reason is that for seismic resistance reasons, the maximum eave height for buildings was set at 100 shaku (a little under 31m) following the 1923 Great Kanto Earthquake that devastated the Tokyo region. This 100-shaku standard was strictly followed until Japan's first skyscraper was built, nearly half a century afterward. For Ando as well, no doubt, a height of a little over 30m is the instinctive default scale of urban architecture. With his impeccable sense of proportion, Ando set the base length at 100m to a side and the height at 34m to create a mass that could be apprehended in its entirety, and then carved into it with the chisel of imagination. To paraphrase Ando's words, he was aiming not to sculpt external symbolism but rather to instill internal dynamism forceful enough to explode into the outside world, and to confer gravity and symbolic power on the structure itself.
Ando employed the road across the north side of the site as the

## 百米见方的"结界"[1]

如果,我被邀请在中国大都市郊外的新城设计一座既是文化中心又将代表新城品格的歌剧院,会怎样呢?当我在距离上海市中心一小时车程的嘉定区下车的时候,我便开始了对这一安藤建筑中尚无先例的作品自问自答式地品评。保利大剧院既要契合城市的规模,又必须唯美地体现新城的象征性,这对于曾经设计世界众多文化设施的安藤忠雄来说,也是首次遇到的难题。

如果单纯就基地的巨大规模而言,似乎可与美国德克萨斯州沃斯堡现代美术馆的设计相比拟,在那里还有路易斯·康的名作金贝尔美术馆[2]。安藤从来不会丢弃他的那股勇于挑战的冲劲,也正是因为这座被称为世界上最美混凝土建筑的金贝尔美术馆激发了他的斗志,从而催生了安藤最为成功的作品——沃斯堡现代美术馆。但是,这一次在嘉定区并没有一座可以视为目标的、里程碑式的文化建筑,安藤将进行一次全新的建筑创作。而当作品完成时,它就会像金贝尔美术馆一样,成为这个地区的名副其实的文化地标。

从建筑本体为底面100m见方的立方体中,我可以窥见安藤的决断。他可以充分利用面向人工湖的宽阔基地,其中并没有必要一定要采用100m这个纯粹的数字。然而,安藤这样做了。在基地的巨大尺度下,他首先有必要确立可以掌控的"自身的宇宙",为此,他选择了100这一数字。在广袤的大地上以"结界"镌刻出"自身的宇宙",这才是设计的第一步。因为,在茫茫大地上,当对空间产生迷乱的时候,基于数学的"边长100m的正方形"植根于大地而生就的"结界",才会真正找到自身的尺度。

因此,我认为,安藤是在有意识地挑战"尺度",从而使自己的作品具有存在感。因此当我看到他的构思草图时,证实了我的推测。因为,与尺度的磨合真实地展现了他的设计灵感。

在草图的正中,描画了一个圆筒形的空洞,与右边一个形似平台的开敞通路堆叠在一起。沿着水平方向横过一条宽宽的蓝色色带,圆筒和平台则用橙色和蓝色的混合色表示。蓝色象征了建筑面向的水面,圆筒形空洞的橙色则象征了木质质感。此外,水边和平台上描画了一些代表访客的小小剪影。

由此,我想到了法国空想建筑师埃特纳·路易斯·布雷[3]设计的牛顿纪念馆及其他未能实现的设计方案。在18世纪末大革命时期描绘的极其夸张的建筑方案中,"夸大狂(来自意大利语megalomania)"布雷为了表现自己的作品可与宇宙的尺度相匹敌,特意在建筑脚下布置了小如细砂的人或景。安藤的草图也让人产生这样的联想。如果说住吉的长屋是凝缩了其个人想法的结晶,那么保利大剧院就是契合了城市尺度并延展了只有安藤特有的构思且足可称为"夸大狂"的作品了。这幅空想建筑师风格的草图充分说明了这一点。

相对于100m见方的底面,安藤设计的立方体高度为34m。这一高度,在日本于建筑师甚至普通人都是熟识宜人的尺度。1923年的关东大地震中,东京受灾严重,从抗震角度出发,自此建筑的挑檐高度就被定为100尺(不到31m)以内,这一规定一直延续到日本最早的超高层建筑建造成功,近半个世纪的时间。安藤也是如此,他一直认为30余米的建筑高度适合于城市的尺度。作为建筑师,安藤忠雄的比例感觉出类拔萃,他把这一底面100m见方高34m的长方体作为一个可全面掌控的体块摆在眼前,展开了鬼斧神工般的想象。借用他的话来讲,这不是在雕琢外在的象征意义,而是在内部运用反复穿插的手法形成韵律感,从而力图让建筑具有力量感

---

1  In Japanese, it is a Buddhist term and it refers to a reserved area for monks to practice Buddhism.
2  Located in the suburb of Fort Worth, Texas, The Kimbell Art Museum was designed by Louis I. Kahn and was finished in 1972. Situated in a spacious and beautiful park, it impresses visitors as gentle, refined and simple. In 1997, Modern Art Museum of Fort Worth was designed and built by Tadao Ando right beside The Kimbell.
3  Étienne-Louis Boullée is the leading architect of French Neoclassicism in 18th century. He is called a visionary architect for most of his designs are never put to practice. Standing on the shoulders of classic architectural principles, he dug into the relationship among light, nature, aesthetics, symbols, the nature of architecture and design language. He broke the bondage of traditions, reeducated people on neoclassicism and made a full display of his foresightedness.

1  日语中原是佛教用语,指"为僧侣的修行而划设一定的区域"。
2  金贝尔美术馆位于美国德克萨斯州沃斯堡的郊区,由路易斯·康设计,于1972年建成。美术馆坐落在一个空旷而景色优美的公园之中,建筑的外观形象处理得娴静、简朴。1997年,安藤忠雄在金贝尔美术馆旁边设计建造了沃斯堡现代美术馆。
3  埃特纳·路易斯·布雷是18世纪法国新古典主义的代表人物,他的大部分设计作品并未落成,因此被人们称之为空想建筑师。在古典建筑设计理念的基础上,他主动探索建筑与光线、自然、美学、象征等因素的关系,以及建筑本质及其设计语言的表达,进而打破了传统建筑的束缚,使人们重新了解和认识新古典主义,并显示出了超越时代的前瞻性。

standard horizontal line for his cuboid structure facing on to the artificial lake. For its interior, he came up with a truly daring spatial configuration.

**Voids Formed by the Force of Modernism**

The flow of visitors into the theatre moves along a vector at a 45° angle to the baseline of the building. In other words, people move along a diagonal in relation to the 100-meter sides. With this decision, it was inevitable that the entrance would be positioned near the corner of the building. From the entrance, the audience moves along a huge, futuristic tubular passageway, diagonal in relation to the building, to the Main Hall. In this hall a stage is positioned with the diagonal line as its central axis, and the audience's line of sight is also in this same diagonal direction.

Generally speaking via the model and sketch, the theatre's interior spatial elements are positioned with the above-described diagonal line as the central axis, and Ando's floor plan shows the square of the building turned at a 45° angle, so that it becomes a diamond. The Main Hall is centred on the stage, with a backstage area and the audience seats to the left and right of it and side-stage areas above and below. This crucifix configuration occupies the left side of the plan, and enclosing it is an array of five tubes. There are three horizontal tubes, the "ground-level amphitheatre tube" (coloured green on the presentation diagram), the "foyer tube" (light blue) and the "connecting tube" (yellow), and two vertical tubes, the "lobby tube" (red) and the "roof-level amphitheatre tube" (gray). Looking at Ando's beautifully coloured three-dimensional spatial configuration model, the Main Hall in a subdued pale blue appears to be silent and static, while the five tubes surrounding it intersect dynamically and seem to be playing the lead role that would usually belong to the Main Hall.

I took this as a visual embodiment of Ando's message that "to opera viewers, the entrance hall and foyer are just as important as the hall." In the hall where the opera is staged, the most critical thing is acoustics-enabling the audience to appreciate the vocal and instrumental performances at their most gorgeous, which is of course an opera house's raison d'etre. However, operas are lengthy and invariably have intermissions, free periods during which the entrance and foyer become open zones for opera lovers to engage in passionate discussions and heated debates over the performance in progress. At opera houses such as Vienna State Opera [4] where the foyer extends outside as a terrace, visitors can revel in the merging of opera, as a form of urban culture, and the actual culture-and-history-rich city that surrounds the theatre, a synergy that elevates the reverberations of the arias into the realm of the sublime. The foyer and entrance hall are places of intrinsic joy and excitement.

By applying his creative sensibilities to these zones, Ando aimed to create such times and spaces of joy in a manner unique to this modernist opera house in a new urban community. The array of tubes described earlier was his unique solution to this problem. Each of the tubes is with only an exoskeleton and no interior architectural elements. The five tubes are boldly deployed and intersect in an effective manner thanks to Ando's scrupulous calculations.

What is the dramatic effect of the Shanghai Grand Poly Theatre's tubes from a visitor's perspective?

Operas are generally staged in the evening, and visitors' first encounter with an opera house is generally a nocturnal one. The Shanghai Poly Grand Theatre, with a glass curtain wall covering its concrete frame, appears luminous, the lighting installed inside the glass making it glow from within. The sash of the curtain wall, designed with Ando's rigorous sense of proportion, is an embodiment of ascetic Modernist aesthetics reminiscent of Ludwig Mies van der Rohe. The apparent luminosity of the building itself, covered with this smooth and refined external membrane, is especially effective in a newly developed suburban district like Jiading where there is little light at night.

If it were truly similar to Mies, Ando's Shanghai Poly Grand Theatre would be merely luminous, but with all four sides penetrated by internal tubes so that great voids open up, there is a dynamic flux not seen in the tranquil geometries of Mies, and a palpable sense that this is what Modernism in the 21st century is all about.

On the east side facing directly on to the artificial lake is a huge opening in the curtain wall, a cross section of the 75° angle intersection of the foyer tube and the ground-level amphitheatre tube. Through this opening the escalator linking the upper floor and the first floor can be seen, and there is a thrill in seeing the bustle of the interior from outside. The shape of the opening, two conjoined ovals, is on the one hand reminiscent of the traditional sukiya architecture characteristic of teahouses, and on the other hand, if Ando were a 1980s postmodernist it might be an intentional reference to the silhouette of a certain

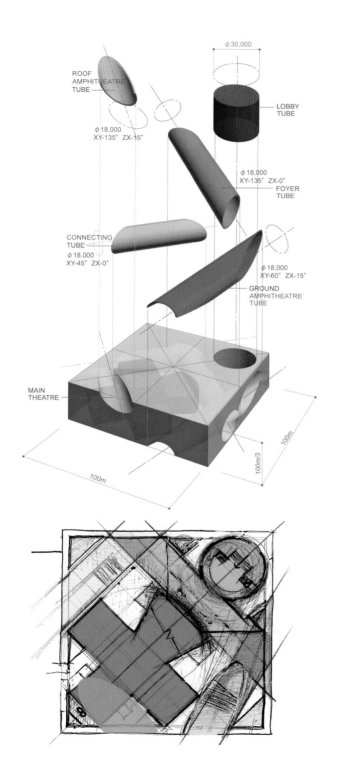

和象征意义。

安藤把基地北面的机动车干道作为水平方向的基准线，将立方体建筑设置于人工湖边上。然后，他开始大胆着手于建筑内部的空间设计。

**被现代主义穿凿的"空洞"**

观赏歌剧的观众流线偏离建筑的基准线45°，边长100m的正方形的对角线便决定了用来引导观众的方向。因此，观众的主要出入口自然而然就布置在了正方形平面的一角。从那里进来的观众朝着正方形的对角线方向行进到主厅，在入口大厅内，舞台以对角线为轴线对称布局，观众的视线依然是沿着对角线的。

如果直接切入形态进行说明的话，剧场空间以对角线为轴线布局。安藤将正方形倾斜45°以绘制平面图。图中的主厅以舞台为中心，呈"十字形"位于平面图中的左半部，其左右分别是后台和观众席，上下则是侧台。围绕着十字形，安藤配置了五个被他称之为"管"的圆筒。这五个分别是水平方向的三个：地面圆形剧场圆筒（表现为图中绿色部分）、休息厅圆筒（浅蓝色部分）、连接圆筒（黄色部分），竖直方向的两个：入口大厅圆筒（红色部分）、屋顶圆形剧场圆筒（灰色部分）。当我看到安藤描画的色彩美丽且表现立体的"空间构成"时，色调稳重的深蓝色主厅静静地守护在一旁，而围绕在周围的五个圆筒却跃动着相互交叉咬合，似乎要取代主厅本来的主要地位。

从中，我可以感受到安藤所主张的"对于身处剧院的观众而言，入口大厅和休息厅是与主厅同等重要的空间"。毫无疑问，上演歌剧的主厅内的席位，对于如何能欣赏到最优美的音响效果至关重要，这几乎可称为歌剧的生命线。在大型歌剧上演必不可少的幕间休息时，观众们离开席位，聚集到入口大厅或休息厅互相畅谈对演奏甚至歌剧的感想，这是他们的放松时间。类似维也纳歌剧院[4]中将休息厅扩展到屋外平台，这类歌剧院作为一种城市文化，与外部的城市文化气息融为一体，演奏的余韵得到了升华。而开启这一幸福时刻的正是休息厅和入口大厅。

安藤随意挥洒着他的造型感，试图创造只有在这个新城的歌剧院才能体会到的成功喜悦。他引入建筑内部空间的便是前述的五个圆筒形。这些圆筒内里只有一层简单的表皮。在安藤的精密计算下，这五个圆筒相互咬合交错，效果出人意料。

那么，从观众的角度看，五个圆筒的设计究竟对保利大剧院的空间构成造成了多大的戏剧效果呢？

歌剧通常在晚间上演，观众首先看到的是暮色中的建筑。夕阳西下，保利大剧院体量巨大，混凝土外覆盖着玻璃幕墙。因着隐藏在玻璃后的照明，它进而变身为闪耀着光芒的"发光体"。安藤依据严格的比例感觉而设计的幕墙格栅，体现了密斯·凡·德罗式的现代禁欲美学。被覆盖在纯净的外表皮下的建筑自身散发着光芒，这种做法使得保利大剧院在缺乏夜间照明的新城中显得更加耀眼夺目。

---

4　Ranked among the most renowned opera houses in the world and regarded as a symbol of Vienna, the Vienna State Opera enjoys the reputation of the centre of the world's opera. Built and superintended by Austrian architects August Sicard von Sicardsburg and Eduard van der Nüll in 1861, the project lasted 8 years. It sits on the Ring Road in the old city of Vienna, and it is originally the Royal opera house.

4　维也纳歌剧院是世界上最著名的歌剧院之一，素有"世界歌剧中心"之称，也是维也纳的主要象征。歌剧院始建于1861年，由奥地利著名建筑师奥古斯特·希卡尔德·冯·希卡尔德伯格和爱德华·范·德·鲁尔设计督造，历时8年完工，坐落在维也纳老城环行大道上，原是皇家宫廷剧院。

Disney character. In any case it has an endearing quality that contrasts with the cool slickness of the building's surface.

On the south wall, to the left of the above-described east wall in Ando's perspective sketch, the bottom half of a slightly off-kilter oval takes a bite out of the roof (this is the roof-level amphitheatre tube), and through this opening the floor of the circular theatre space on the roof can be seen. At ground level, the dome-shaped upper part of an oval cross-section, the narrowed-down tip of the ground-level amphitheatre tube, creates another opening, which draws attention to the ground-level outdoor theatre space. The slices taken out of both top and bottom are a feat of architectural muscle that disrupts the calm of Mies-like modernism.

On the north wall is the void of the "ground-level tube," and on the west wall that of the "connecting tube," both enormous openings extending up to the fifth floor of the building. Now, let us proceed into the building interior.

## A Modernist Interior Born out of Interplay and Collision

The entrance that leads visitors in is positioned at the northeast corner of the prismatic structure. Upon entering, one is immediately faced with the curved wall of the soaring vertical "lobby tube." Overall, the foyer, passageways, and other public areas of the Shanghai Poly Grand Theatre are surfaced in cast aluminum blocks that at first appear to be squared-off timbers. The curved surface of the lobby tube exterior and interior are treated in the same way, and visitors encounter and are swept into what seems to be an immense wood cylinder. Proceeding into the lobby and looking upward, one sees a glass roof that lets in natural light, with a square plate mounted a short distance inside the circle of the roof to mitigate the blinding light of midday. In the daytime, the light that penetrates through the spaces between the circular roof edges and the square shield is still bright, and makes visitors feel they are at the bottom of a luminous well. Looking upward, they also see another tube bisecting the lobby tube horizontally at the fourth-floor level.

Having experienced the lobby tube space, visitors ascend one of the pair of stairways positioned to the left and right, elements that are practically de rigueur in a work by Ando, and ascend to the foyer level. There they find an interior not quite like anything encountered before, thanks to the intersection of tubes, as there is an enormous round hole in the back of the vertical lobby tube where it intersects with the horizontal foyer tube. As described earlier, the tubes are voids with nothing but an external shell slicing off their tubular interiors, and where their curved surfaces intersect, there are amazing joints that enthrall the viewer.

What enthralls is the variety of oval openings created when tubes intersect one another diagonally. Some tubes are tilted or twisted, and when tubes collide in the spaces leading into the foyer, it generates wall surfaces unlike anything we have seen before.

The ground-level amphitheatre tube is tilted upward at a 15-degree angle to the horizontal axis of the building. When this intersects with the truly horizontal foyer tube, it creates a number of cross-sectional ovals. When it intersects with the vertical lobby tube at the same 15-degree upward-tilted angle, it creates another series of curved surfaces. These curved surfaces are in turn slashed open by ovoid voids, and the visitor is beckoned into an unprecedented spatial experience. The remaining two tubes, the roof-level amphitheatre tube and the connecting tube, also intersect, and the overall outcome is a futuristic interior space born out of interplay and collision between geometric forms, with a dynamism that belies the simplicity of the square framework 100m to a side.

In the process of designing this building, the computer certainly plays a major role, for instance in determining the method of transporting wood-grain-printed sections of cast aluminum material to their places on the site. However, I certainly would not say this hall and foyer are examples of architectural design strongly aided by computer graphics. Ando's inspiration to introduce these five cylindrical voids into the building interior is one fruit of his many years as a modernist composing structures out of solid and planar geometric figures. I highly doubt that Ando envisioned relying on computers in conceiving the intersections of these five tubes and the challenging task of working out their detailing. I believe that as a modernist architect, just like Mies, Ando quite naturally and simply inserted five enormous tubal voids into a geometric solid 100m square at its base and 34m high, and let the intersections between these tubes exert their dynamic effect. Ando's unrestrained admiration for the "blood, sweat, and tears" of Chinese builders on this project comes out of his gratitude for their unstinting efforts to realize his architectural concepts. In particular, he sings the praises of engineers and builders whose expertise enabled the technically and visually elaborate joining of arc and oval using cast aluminum in the foyer and entrance hall. The enthusiasm and passionate embracing of difficult challenges that is becoming increasingly rare on Japanese construction sites was undoubtedly present in full force on the Shanghai Poly Grand Theatre project. It is evident that the builders here took on the architect's wishes as their own personal mission, and rather than expecting to achieve a solution through digital means, solved each challenge in an analog fashion on site. Obviously deeply moved by their performance, Ando says, "In Japan, Japanese builders would have given up on doing this right at the start. The Chinese builders, however, resolved each difficulty that came up, and managed to build something that exists nowhere else in the world." The powerful impact of the Shanghai Poly Grand Theatre, which began with just the concept of a 100-meter square demarcated in a vaguely defined space in a newly developed suburb, has taken shape brilliantly and turned the outskirts of Shanghai into the site of a new

不，如果是密斯的建筑或仅限于此吧。安藤的保利大剧院，却因为"圆筒"贯穿建筑内部而在四面的外墙上均出现了巨大的空洞，带来了密斯的静谧建筑所不能比拟的动势，让人感受到了21世纪的现代建筑的存在。

正对着人工湖的建筑东立面幕墙上，与水平方向呈75°交叉的两个休息厅圆筒和地面圆形剧场圆筒以相连接的方式挖开了巨大的空洞。人们用一个非常有趣的方式从空洞中看见连通一层和二层的扶手电梯，甚至是建筑内部的喧哗。两个椭圆形相连的空洞，它的样子似乎是日本数寄屋建筑中的窗，又像是20世纪80年代后现代时期的迪斯尼漫画中的剪影。不管怎样，它的可爱与建筑的冰冷外皮形成了对比。

在它的左手边是建筑的南立面，玻璃幕墙的挑檐部分中，稍许扭曲的椭圆形的下半部分被屋顶圆形剧场圆筒削掉，从那里我们可以看见屋顶圆形剧场的地面。甚至在脚下开凿了一个部分呈椭圆形状的圆洞，它就在稍稍扭曲圆形剧场的最顶端，就好像一个郊外剧场。挑檐和底部都相应地被切割了一部分，安藤也许正是用他的这种设计思想突破了密斯的建筑美学。

我们还可以从北立面的圆形圆筒以及西立面的连接圆筒中看见高达五层的底层空间的空洞，人们可以沿着它们横向地在空间中漫步。

## 交错及冲突，内部空间的活力

作为观众主动线的入口位于长方体的东北角上。由此进入以后，眼前正对的是竖向耸立的入口大厅圆筒的曲面墙。保利大剧院的休息厅、通道等公共交通部分的室内几乎都采用了看似角钢的铝质型材。入口大厅圆筒的外部曲面与内部空洞采用同样的做法，观众直面"木质圆筒"而逐步进入其中。抬头仰望，可见圆筒的上部是可以采光的玻璃屋顶，为了避免正午的强烈日照，在靠近玻璃屋顶的地方设计了一个与入口大厅圆筒的正圆形内接的正方形金属板。虽然在日间，正圆与正方形之间的空隙中透射下来的光线仍然非常强烈。观众们也因此似乎置身于"光的井底"。而且，还能看到另一个圆筒在四层高的空间横穿入口大厅圆筒。

体验过入口大厅圆筒空间感觉的观众，顺着位于圆筒中央呈左右布局（几乎是安藤作品中的定式）的楼梯向上可以抵达休息厅层。在那里观众将要看到"从未见过的室内设计"，竖直方向的入口大厅圆筒的尽头与水平方向的休息厅圆筒相交错。至此观众看到的"圆筒"作为"空洞"，既截然切开内部空间，曲面间又相互交错，都使得观众在视觉上感觉纷扰扰又不可思议。

而生成这些不可思议的视觉效果的是一些椭圆形断面，它们在建筑空间中被众多圆筒斜向切断而显得多种多样。这些断面时而倾斜时而扭曲，组合在一起后就形成了"从未见过"的休息厅空间。

地面圆形剧场的圆筒相对水平的坐标轴向上倾斜了15°，这一向上的圆筒与水平的休息厅圆筒相交错而生成了一系列的椭圆形断面，并与竖直方向的入口大厅圆筒曲面以向上倾斜15°相遇形成了连续曲面。安藤将这一连续曲面切成空洞，让驻足在休息厅的人们"体验未知的空间"。另外在

余下的两个屋顶圆形剧场和连接的圆筒中，两个圆筒之间的连接使得底面100m见方的立方体中充满了生动的几何体相冲突的内部空间。

在将具有自然木质效果的铝制型材安装在圆筒内壁时，电脑制图发挥了很大作用。然而，入口大厅和休息厅的造型却并非依靠电脑制图。

将五个圆筒植入内部空间，这一构思基于安藤长年对采用几何形体构筑建筑的摸索积累。通过将五个圆筒交错布局，实现高难度的细部做法，难以想象这些都是仅依赖于电脑的帮助。我以为，安藤作为现代建筑家，至少正如密斯等一样，不过是很自然率性地将五个圆筒作为几何体直接插入长、宽100m，高34m的立方体中，这些圆筒的交错引发了令人期待的跃动感。

安藤不吝去赞美中国的施工方付出的"血汗与泪"，更是为了感谢他们不遗余力地去实现他的纯粹造型。他由衷地赞美了休息厅和入口大厅内部的圆弧与椭圆弧交错节点的铝制型材的使用、现场进行精密施工的技术人员和施工人员的高水准。毫无疑问，在保利大剧院中，充满了在日本国内施工现场日渐稀少的挑战难题的意愿和激情。这些，不是基于数字计算得出的预测结果，而是现场的施工人员因为建筑师们的"这里我要这样做"的意图，而在现场反复试验修正错误的守旧做法。安藤在回顾这项工程的时候感慨道："在日本，一个想法提出伊始就会被认为是不可能的，而在中国，这些难题会在现场逐一被解决，从而建造出世界上独一无二的建筑。"身处空旷的新城中，100m见方的立方体的存在感，成就了安藤迄今为止未曾有过的大型建筑。我衷心地祝愿保利大剧院作为日中同时代的文化交流一环顺利竣工。

## 由现代的"坦比哀多礼拜堂"[5]开始

1,600座席的主厅的内装修采用了合成板材。这既是考虑到音响效果，

---

[5] Located in Munto Leo in Rome, the Tempietto is designed by Donato Bramante and is representative of Italian Renaissance. When religion is concerned, it is where St Peter died. In building, it is the leading architecture during the peak of Italian Renaissance (1502). Quite a few later buildings, like Saint Peter's Basilica, Pantheon in Paris and even United States Capitol, have been inspired by it.

[5] 坦比哀多礼拜堂由多纳特·伯拉孟特设计，位于罗马的蒙托利欧，是意大利文艺复兴时期的著名建筑。宗教上，它是圣彼得殉难处。建筑上，它是文艺复兴鼎盛时期的纲领性作品（1502年），后期的很多建筑，比如圣彼得大教堂、巴黎的先贤祠，甚至美国的国会大厦都汲取了它的设计灵感。

Ando architectural masterpiece. This is cause for celebration in terms of its facilitation of contemporary cultural exchange between Japan and China as well.

## A Journey That Began With a Modernist Tempietto [5]

The interior of the main hall, which seats 1,600, is finished in butcher's-block panels. The walls on either side of the seats are an orderly sequence of shallow curved surfaces, no doubt designed with acoustic properties in mind. The curving lines described by the panels have protrusions and indentations that heighten the sense of mass. The slightly orange-tinted wood grain creates a pleasant and comforting atmosphere that makes someone seated in the audience feel their entire body is enveloped in and harmonizing with the "breathing" of the wood surface.

A gradual slope ascends back from the stage on the ground floor, and curved areas of balcony seating thrust out on either side. The main stage proscenium, which can be seen from the audience across the orchestra pit, has a width of about 25m, and the side-stages on either side have widths of about 20m each. Both main and side stages have a depth of approximately 20m, and there is a revolving stage behind the main stage, a necessity for rapid changes of scenery during the staging of operas.

In all of China, there are 37 opera houses with this sort of configuration, seating a total of 71,126 and they are run with the expertise of the Beijing Poly Theatre Management Co., Ltd. As Ando himself notes, he has designed many museums around the world, but this was his first time designing an opera house, and in terms of logistics it was very much a collaboration with the Poly group.

The Shanghai Poly Grand Theatre also has a Multi-Purpose Hall on the fourth floor, as well as a circular outdoor theatre area where the ground-level amphitheatre tube juts through the exterior wall to form the proscenium, and another circular outdoor theatre on the roof where a tube slices through the south façade at the top. There is a gallery space to the side of the foyer on the first floor. All together, the building is expected to serve as a composite facility forming the core of the Jiading district's cultural life.

The interior décor of the Multi-Purpose Hall is smoothly sophisticated, in contrast to the warm wood grain of the Main Hall. Meanwhile, the circular outdoor theatre on the ground level protrudes out from the building frame, appearing to float on the surface of a man-made pond that surrounds the building. From the viewpoint of audience members seated on risers inside the building, the backdrop to the performers is an unbroken expanse of water, with this man-made pond merging into the larger artificial lake around the structure.

To give shape to these multiple functions, Ando mobilized the entire expressive lexicon he has amassed thus far, and added a daring new piece of vocabulary to it with the intersections of enormous tubes. It is clear evidence that his intrinsic fighting spirit burns as brightly now that he is an internationally acclaimed architect as it did during the early days of struggle.

Let us turn once again to the issue of scale, which was discussed at the beginning of this review.

Ando's first famous project on the international architectural scene, Sumiyoshi Row House, was a small one, with a façade 3.6m wide facing the street, and a depth of 12.6m. That this truly compact urban residence became an internationally recognized architectural landmark is nothing short of a miracle of 20th-century architecture. In that sense, the Sumiyoshi Row House is the modernist equivalent of the Tempietto in Italy.

The Tempietto, a tiny domed martyrium (tomb) forming part of the church of San Pietro in Montorio, Rome, was designed by Donato Bramante in the early 16th century, and set a defining standard for Renaissance architecture. Its stature ranks with that of another, much larger San Pietro, i.e. St. Peter's Basilica in the nearby Vatican, and this feat of compact architecture from 500 years ago is a fabled chapter in architectural history that resembles that of the tiny, extraordinarily acclaimed Sumiyoshi Row House.

The trajectory of Tadao Ando's career began with the recognition he garnered as a minimalist, and as the nature of his commissions evolved from small residences to commercial facilities to public cultural landmarks, the scale at which he conceives structures appears to have expanded accordingly. However, in crowded Japan there is a limit to expansion of scale, and his only work of truly epic proportions is the multi-purpose resort complex Awaji Yumebutai. Outside Japan, however, he has designed a great number of large-scale cultural facilities, among which the Modern Art Museum of Fort Worth shines as a masterpiece.

The journey that began with a 3.6-meter-wide façade has reached a new milestone with the Shanghai Poly Grand Theatre, an enormous piece of solid geometry with a building frame 100m to a side. It is a powerful answer to the rising expectations of an international architectural scene that has been tracking Ando's progress.

As I mentioned earlier, this new landmark also deserves to be applauded by both Japan and China as a testament to the two countries' contemporary cultural relationship. It will also serve to raise even further the already high expectations for other Ando works scheduled for completion in various locations throughout China.

(Kazukiyo Matsuba  Architectural critic/ Professor of the Musashino Art University)

也是因为座席两侧的墙壁上规则分布的微曲面施工的需要。具有凹凸感的合成板材叠砌出的水平线更加增强了曲面墙壁的质感。木纹稍稍呈桔色，身在其中不禁使人感到树的气息。

座席的一层部分随着阶梯的升高形成和缓的坡度，两侧还设置了围合式的曲面楼座。从座席越过乐池看去，舞台宽约25m，两侧的侧台各自也宽约20m。主台侧台各自进深达20m，另外歌剧院常设的后台也被设计为舞台的回路，这有助于演出时快速地转换背景。

这样的剧场，得益于北京保利剧院管理有限公司对旗下37个剧场共71,126个座席数的管理及运营能力。正如安藤自身所说的，他虽然曾着手于世界上多个美术馆的设计，但却是首次设计歌剧院，此次他实现了与保利集团的携手合作。

另外，保利大剧院还包括了四层的多功能厅；室外的"地面圆形剧场圆筒"与外墙面形成的室外"空洞"舞台；在屋顶的圆筒与南立面所形成的另一个室外圆形剧场。在一层的休息厅一侧还设置了展示空间，它将被展示为新城中的复合文化设施中心。

多功能厅的内装修，是与主厅的木质纹理相对照的现代的简约风格。地面层的室外圆形剧场的主舞台，被设计为围合建筑本体的形式，好似漂浮在浅浅的人工湖中。观众从建筑本体内部的楼梯位置望去，可以看见舞台背后人工池与人工湖的广阔水面相接。

对应建筑的复杂功能，安藤几乎使用了他迄今为止的所有建筑语汇，同时他也挑战了运用"圆筒"以形成空间碰撞的表达形式。即便现在他已身为世界级建筑大师，但他作为"屡败屡战的建筑家"依然斗志高昂。

在这里要返回去说一说开篇提到的尺度。

安藤初次登上国际建筑舞台的成名作住吉的长屋，体量狭长，开间2间，进深7间，也就是说建筑面向道路仅宽3.6m、进深12.6m。这个真正袖珍的城市住宅，竟然席卷世界建筑界而无人不知，可谓是20世纪的建筑奇迹。

在这个意义上，住吉的长屋便是漫步于现代的坦比哀多礼拜堂。

16世纪初，根据多纳特·伯拉孟特（1444~1514）的设计而建造的极小的穹顶式建筑坦比哀多礼拜堂，有着文艺复兴建筑的典型风格，它被与近旁的梵蒂冈圣彼得大教堂（15世纪重建）并肩比拟，可谓建筑史上的佳话，而这又与对住吉的长屋的破格评论有异曲同工之妙。

安藤忠雄的建筑活动，随着他的建筑项目由小型住宅逐渐向商业建筑、公共文化建筑等扩展，其自身创造的建筑规模及尺度也渐渐扩大。然而，在国土面积狭小的日本，尺度的扩大可谓有限，最大也就是淡路梦舞台。另一方面，他在日本以外却设计了众多的大规模文化建筑，诞生了如沃斯堡现代美术馆等佳作。

由3.6m开始的建筑创作之旅，到上海保利大剧院扩身为仅建筑本体便为100m见方的立方体，安藤借用这一作品回应了世界建筑界对于他的建筑轨迹的期待。

虽言之过早，但从中日建筑交流的观点出发，这个作品值得两国为之大声喝彩。而且，毫无疑问，这一建筑的成功必将使人们更加期待安藤忠雄在中国国内即将相继竣工的其他作品。

［松叶一清　建筑评论家／武藏野美术大学教授］

P97: The sketch of the initial shape studies
P98: The enriched gradation of open space under the night screen
P99: Top, The diagram of the axis angle and diametre location from the five cylinders crisscrossed into the cube; Bottom, The sketch plan of Opera space
P101: Multiple cylinders crisscrossing with each other makes the interior space marvelous sections
P103: Auditorium exudes an aura of soft and elegant color

97页：初期形体研究手绘草图
98页：夜幕中层次丰富的圆筒开口空间
99页：上，五个圆筒插入长方体的轴线角度和直径位置示意图；下，剧院空间手绘平面布置图
101页：剧院内部空间被众多圆筒相互交错切断形成不可思议的椭圆形断面
103页：散发着柔和高雅色泽的观众厅

PP104-105: Light illuminates the entrance hall's horizontally-aligned wood grille, whose contrast with the fair-faced concrete and stairs creates layers of meaning in the space
PP106-107: Shadows dance upon the curving walls

104-105 页：大厅天井的采光照亮了整个横向木格栅饰面墙体，搭配上清水混凝土墙面以及阶梯，创造出丰富的层次感
106-107 页：光影在弧形墙壁上跳跃着

PP108-109: Positions of the aerial corridors relative to the cylindrical space
PP110-111: Looking up, the cylinder diffuses light through its apertures, with the various light effects forming different "smiling faces"

108-109 页：空中廊道与圆筒的相对关系
110-111 页：在圆筒空间内抬头仰望，阳光透过开口洒向圆筒四周，随着光线不同的变化，形成了一张张不同的"笑脸"

P112: View of the cylindrical space—even simple geometric forms can create lively and complex spaces
P113: The crisscrossing of cylindrical spaces at the entrance hall and foyer produces a perfect arc-shaped opening
PP114-117: The rich layering of the foyer is another stage for spectators
PP118-119: The intersection of cylindrical space with one of the façades creates a semi-outdoor terrace
PP120-121: The crisscrossing of cylindrical forms creates arcs with different curvatures, resulting in a variety of apertures

112 页：俯视整个圆筒空间，即使简单的几何元素，也可以创造出复杂且生动的空间
113 页：入口大厅的圆筒空间与休息厅的圆筒空间交错后形成了完美的弧线形开口
114-117 页：层次丰富的休息厅空间，是留给观众的另一个舞台
118-119 页：圆筒空间与外立面的交错，创造出了半室外的露台空间
120-121 页：各个圆筒相互交错而成的不同曲度的弧线，让开口形式变得更加丰富

PP122-123: Visitors can enjoy a diverse range of outdoor and indoor space
PP124-125: The stage of the amphitheatre, with its direct access to Yuanxiang Lake, is not only suitable for the performance of classics, but also for holding civic activities
PP126-127: Voids in the cylindrical spaces allow for architectural dialogue between different levels
PP128: The vertically aligned fair-faced concrete, displayed in the foreground, strikes a sharp contrast with the curved wooden panel surface in the distance

122-123 页：从室内到室外，观众可以体验到各种不同的空间
124-125 页：半室外剧场舞台直接通向远香湖，这里既可以上演经典剧目，又可以成为市民举办活动的场所
126-127 页：圆筒空间之间还有挑空，让不同层高空间之间进行对话
128 页：近处垂直的清水混凝土与远处曲线的木格栅饰板，形成强烈的对比

P129: Top, Visitors are able to enjoy the structure's spatial uniqueness as they ride escalator traversing multiple cylindrical spaces; Bottom, Corridors and stairways are defined by the fair-faced concrete wall
PP130-131: The completed auditorium exudes an aura of musical elegance
PP132-133: The auditorium wall is composed of laminated pinewood, set in a staggered, convex design
PP134-135: Once the curtains are up, the audience will be treated to an audio-visual feast

129 页：上，作为连接不同空间的自动扶梯在圆筒空间内穿梭，让参观者即使穿越时，也能体验有趣的空间感；下，圆筒空间中的清水混凝土墙定义出廊道以及阶梯
130-131 页：建成后的观众厅散发出如同乐器般柔和高雅的色泽
132-133 页：观众厅的墙面采用凹凸错落的松木集成材
134-135 页：当帷幕拉开时，观众们将迎接一场视听盛宴

# Drawings
图纸

Master plan (scale:1/1,600)
总平面图（比例：1/1,600）

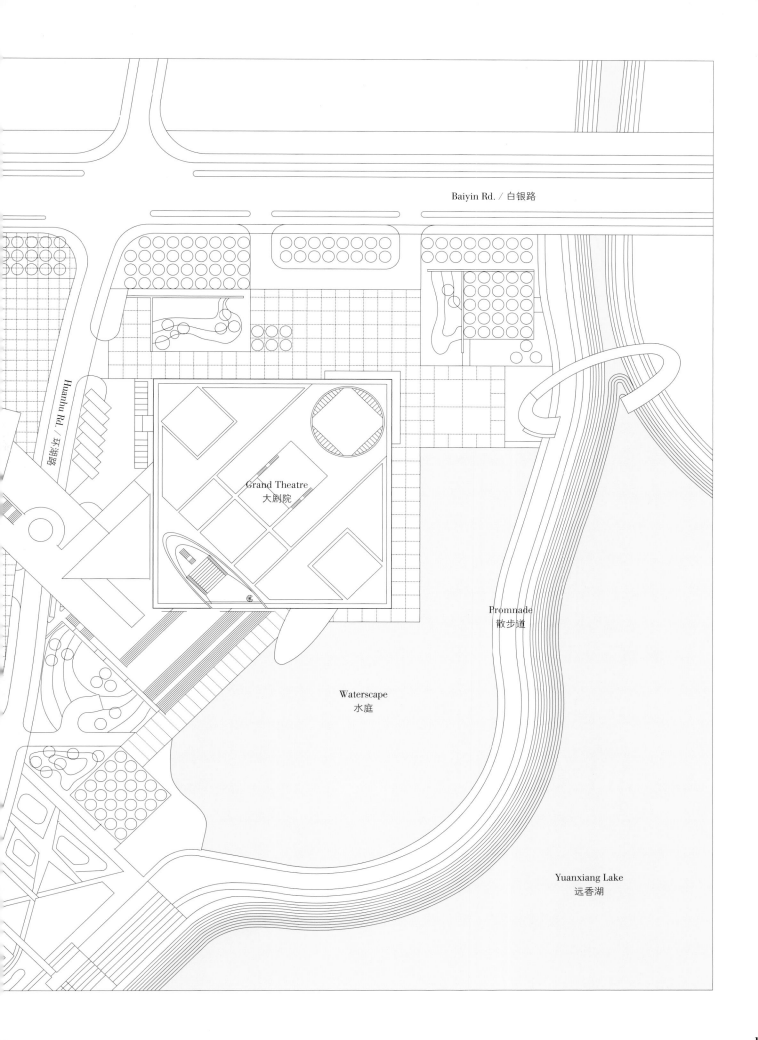

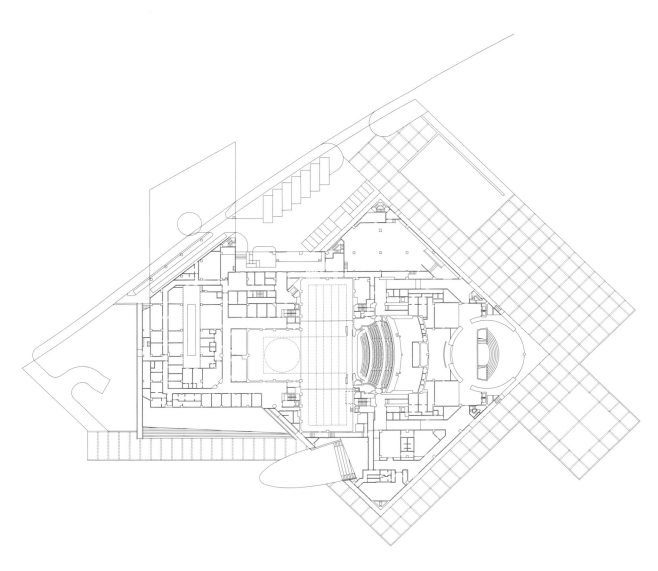

1F Plan (scale:1/1,600) / 一层平面图（比例：1/1,600）

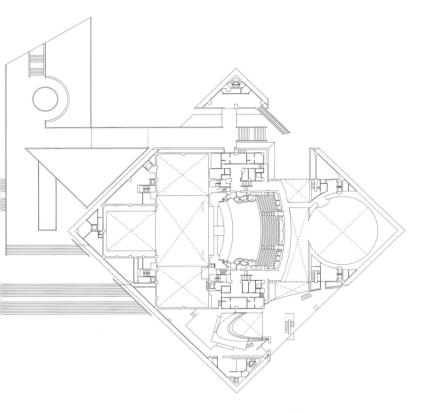

3F Plan (scale:1/1,600) / 三层平面图（比例：1/1,600）

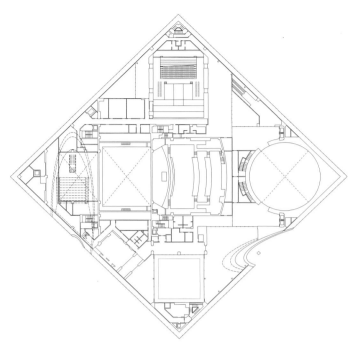

4F Plan (scale:1/1,600) / 四层平面图（比例：1/1,600）

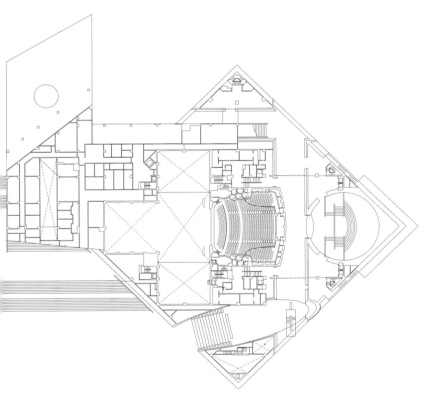

2F Plan (scale:1/1,600) / 二层平面图（比例：1/1,600）

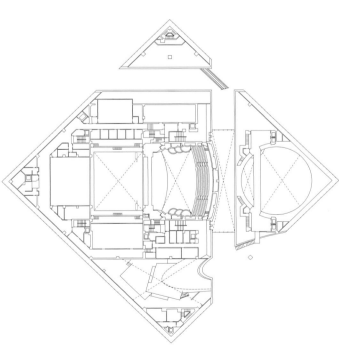

5F Plan (scale:1/1,600) / 五层平面图（比例：1/1,600）

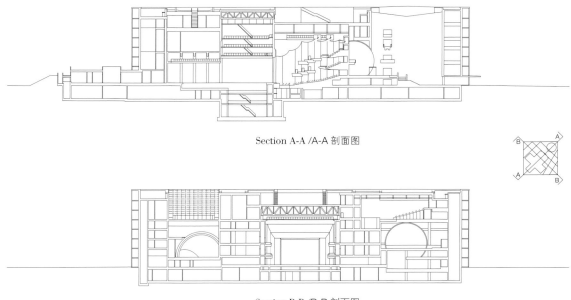

Section A-A / A-A 剖面图

Section B-B / B-B 剖面图

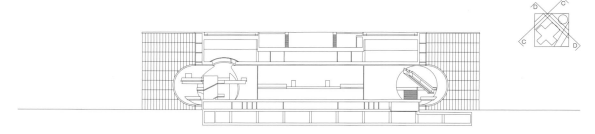

Section C-C / C-C 剖面图

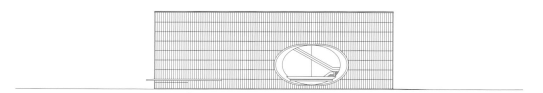

Section D-D (scale:1/1,600) / D-D 剖面图（比例：1/1,600）

North elevation / 北立面图

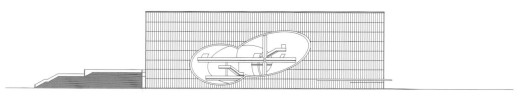

East elevation (scale:1/1,600) / 东立面图（比例：1/1,600）

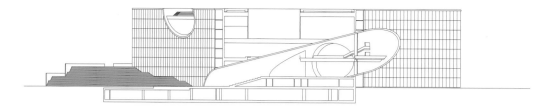

Section E-E / E-E 剖面图

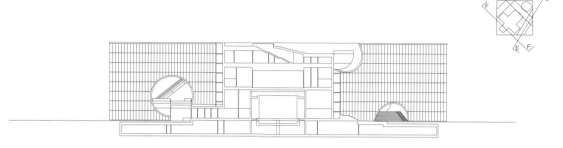

Section F-F / F-F 剖面图

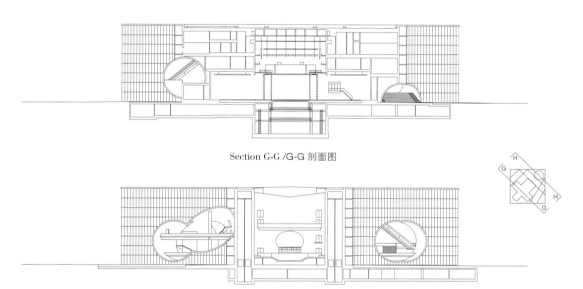

Section G-G / G-G 剖面图

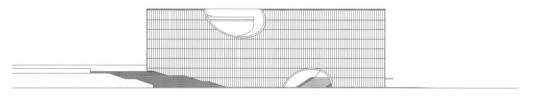

Section H-H (scale:1/1,600) / H-H 剖面图（比例：1/1,600）

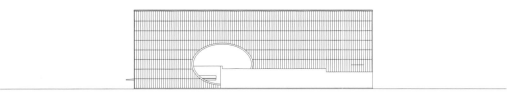

South elevation / 南立面图

West elevation (scale:1/1,600) / 西立面图（比例：1/1,600）

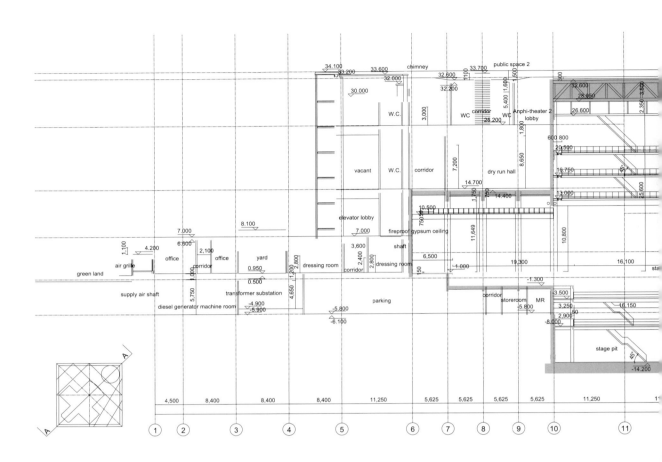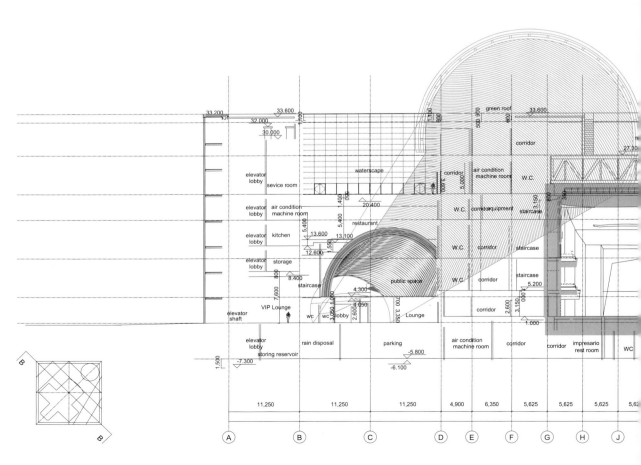

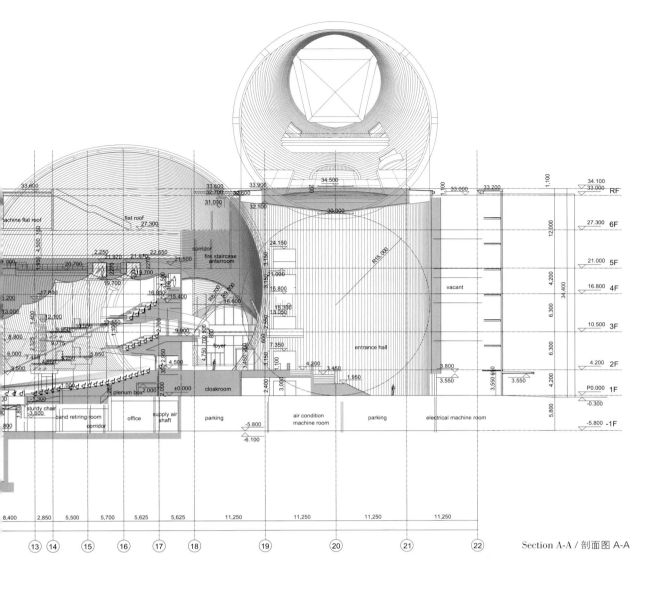

Section A-A / 剖面图 A-A

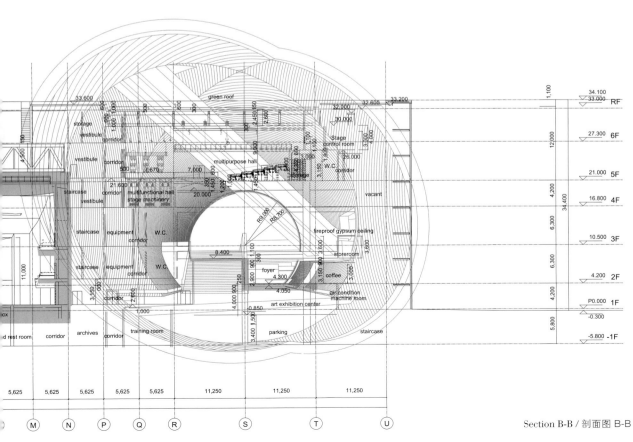

Section B-B / 剖面图 B-B

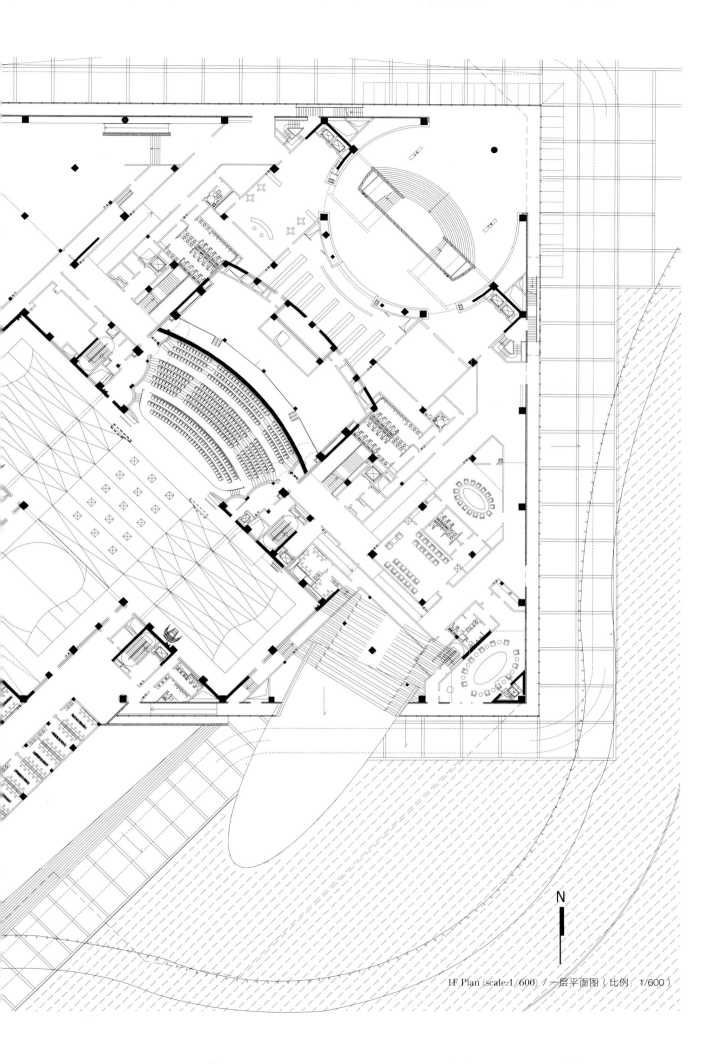

1F Plan (scale:1/600) / 一层平面图（比例：1/600）

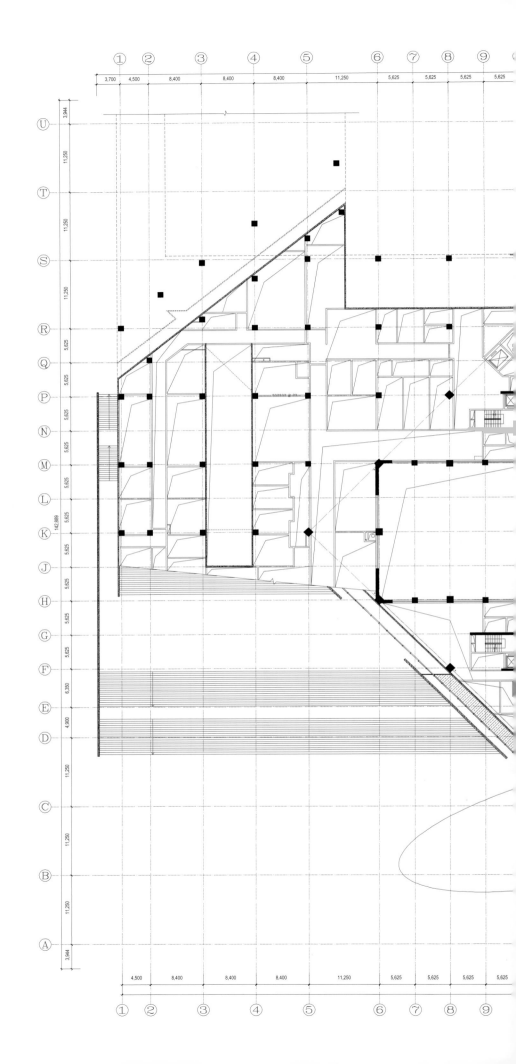

2F Plan (scale:1/600) / 二层平面图（比例：1/600）

Floor area: 3,062m²
Outdoor space area: 910m²
Curtain wall area: 405m²

本层建筑面积：3,062m²
室外空间面积：910m²
幕墙面积：405m²

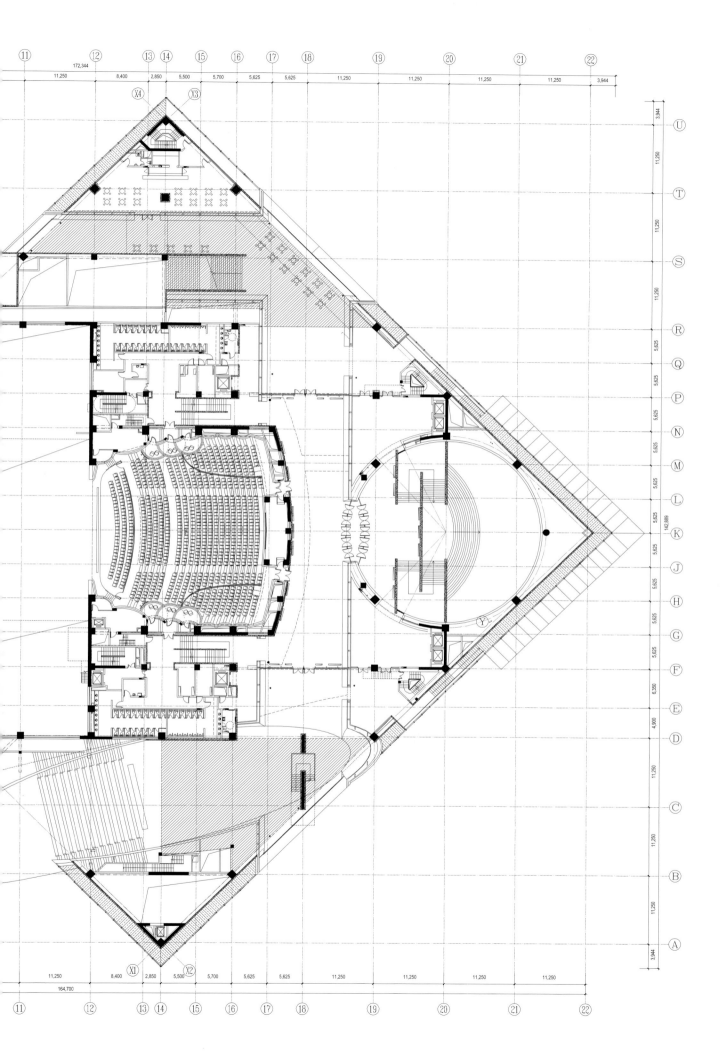

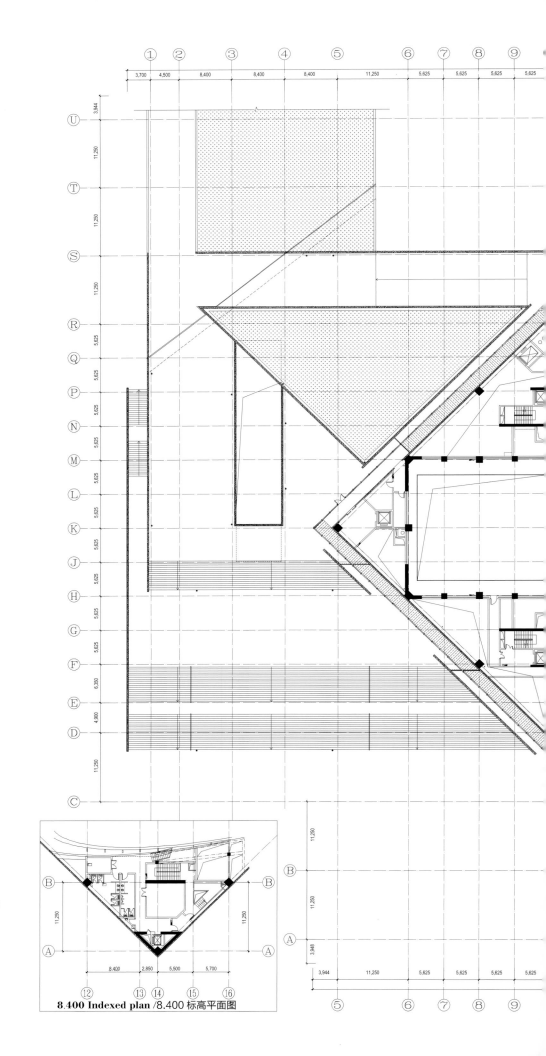

3F Plan (scale:1/600) / 三层平面图（比例：1/600）

Floor area: 3,157m² (Calculate the plot ratio)
Outdoor space area: 433m²
Curtain wall area: 584m²

本层建筑面积：3,157m²（计容积率）
室外空间面积：433m²
幕墙面积：584m²

8.400 Indexed plan / 8.400 标高平面图

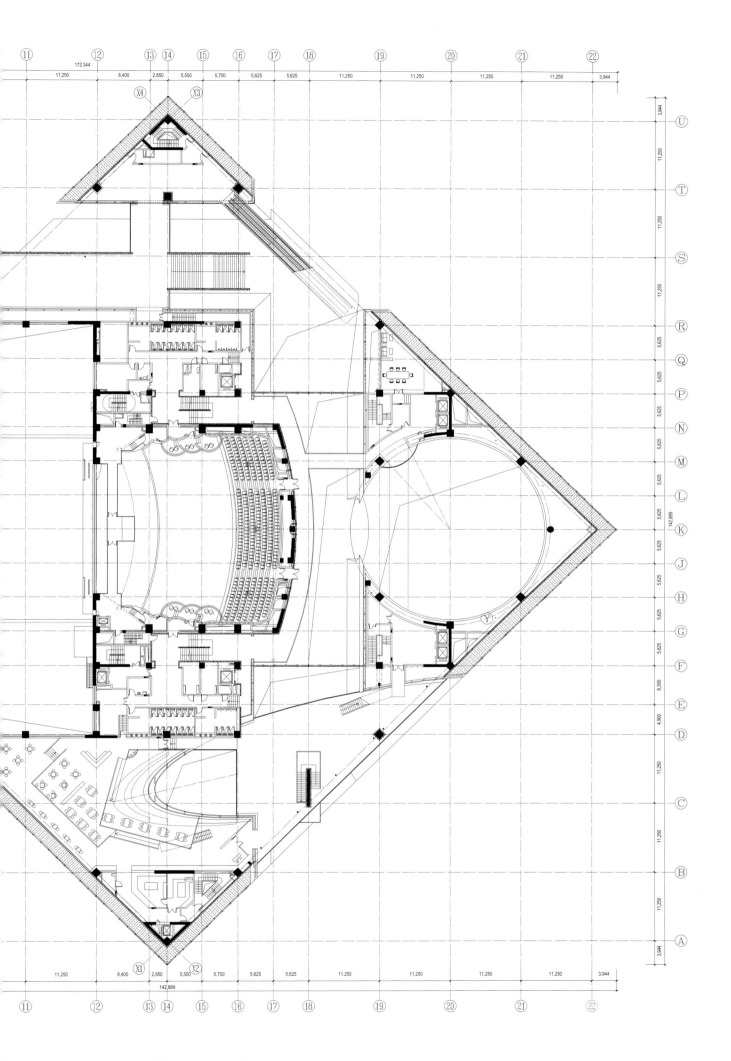

4F Plan (scale:1/600) / 四层平面图（比例：1/600）

Floor area: 3,850m²
Curtain wall area: 666m²

本层建筑面积：3,850m²
幕墙面积：666m²

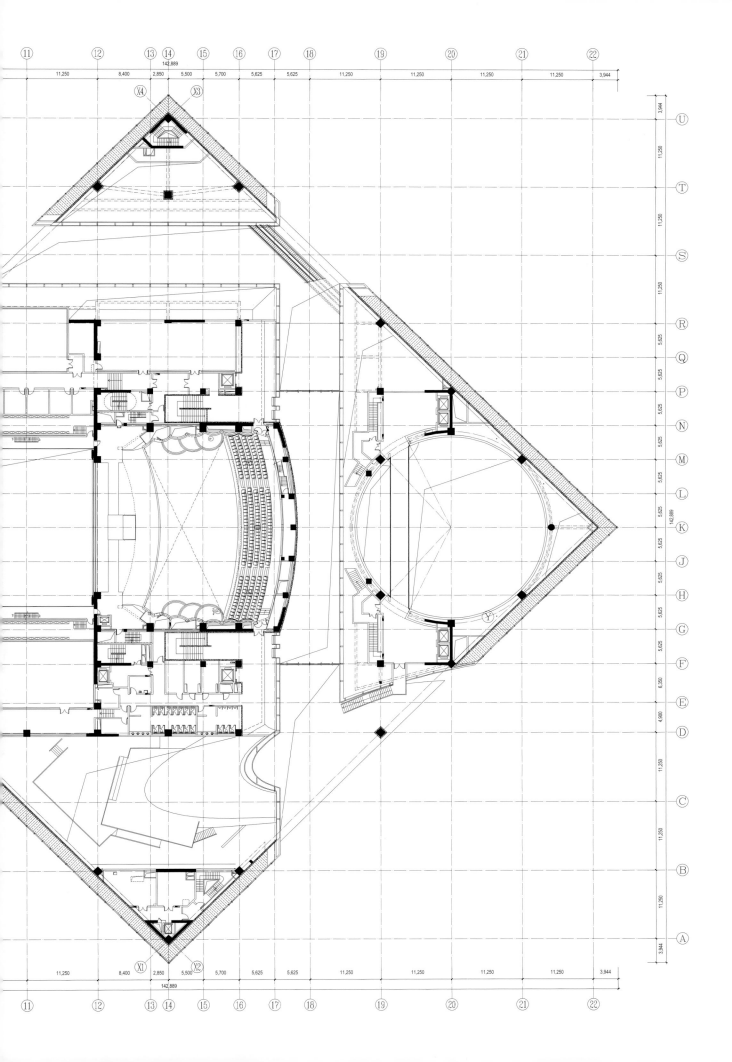

CHAPTER 2 | 第 2 章

# Features
特点构成

**Fair-faced Concrete**
清水混凝土

**Curtain Wall**
幕墙

**Cylinder Nodes**
圆筒节点

**Interior Decoration of the Auditorium**
观众厅室内装饰

**Lighting**
照明

**Theatre Acoustics**
剧场声学

# Fair-faced Concrete
清水混凝土

The concrete pouring process is one of the most common techniques employed in modern construction, yet involves a great sense of fear as one cannot ascertain the result until the removal of the moulding. All of the following elements factor into the success or failure of fair-faced concrete: the design and segmentation of the moulding, the water-to-cement ratio, aggregate type, variations in rebar thickness, management of the pouring process, personnel training, and product quality assurance.

混凝土浇筑工艺是代表现代建筑最为普通的施工技术，但是这一最为普通的技术中，却隐藏着不到拆除模板的那一刻是无法知晓浇筑成功与否的那种恐惧感。立面模板分割的设计、水和水泥的比例、骨材的种类、钢筋的粗细间隔、模板的式样、浇筑的组织管理、人员的培训、成品的保护等等，都关系着清水混凝土最终的成败。

P154: Photo of the fair-faced concrete construction in process
P155: The construction team discusses and makes onsite adjustments to the fair-faced concrete casting process

154 页：后台区域的混凝土现场施工过程照
155 页：组图为施工团队在施工现场对混凝土浇筑工艺的研讨和修正

P156: Construction workers check the concrete after the moulding is removed
P157: The construction site of the cylindrical spaces-in concrete

156 页：施工过程中，施工人员在查看脱模后的混凝土效果
157 页：圆筒空间混凝土施工现场

Fair-faced concrete is the lifeline of Ando's architecture, hence it comes as no surprise that Ando Architect & Associates is well-renowned, even to the point for perfection, for its high standards and requirements in the medium and its construction. In order to execute fair-faced concrete that satisfies Ando's standards, a mould is derived from detailed construction drawings, which includes the opening positions and sizes of all the switches, lighting, and electrical equipment. Elements of the concrete wall, such as structural openings, seams, and joints, are constantly redesigned and readjusted during the construction process. Three-dimensional models are needed to determine the precise dimensions of the concrete wall, which are influenced heavily by its curvature. Prior to the actual construction, dozens of tests were conducted on the fair-faced concrete. Thanks to the meticulous design, the around-the-clock construction process, and strict management, in the end the Poly Grand Theatre was built using 36,500m$^2$ of high-quality fair-faced concrete.

清水混凝土墙体可说是安藤建筑的生命线，所以对于清水混凝土施工，安藤事务所一直以高标准严要求著称于世，可以说有些地方甚至已经到了苛求的程度。为了在中国实现真正合格的"安藤清水混凝土"，设计师在建筑施工图的基础上专门绘制了清水混凝土模板图，在模板图上标注了所有的开关、照明、电器等相关设备的开口位置及具体尺寸。拉杆孔、禅缝、施工缝等构成墙体的建筑元素都必须经过再三的设计和调整。一些清水弧墙由于有一定的曲率，所以需要采用制作三维模型的办法来最终确定模板的细微尺寸。除此之外，在正式施工前，施工团队做了十几次清水混凝土的试验。设计上的精益求精，再加上现场日以继夜的施工和严格的管理，最终成就了保利大剧院近 3.65 万 m$^2$ 的高品质清水混凝土。

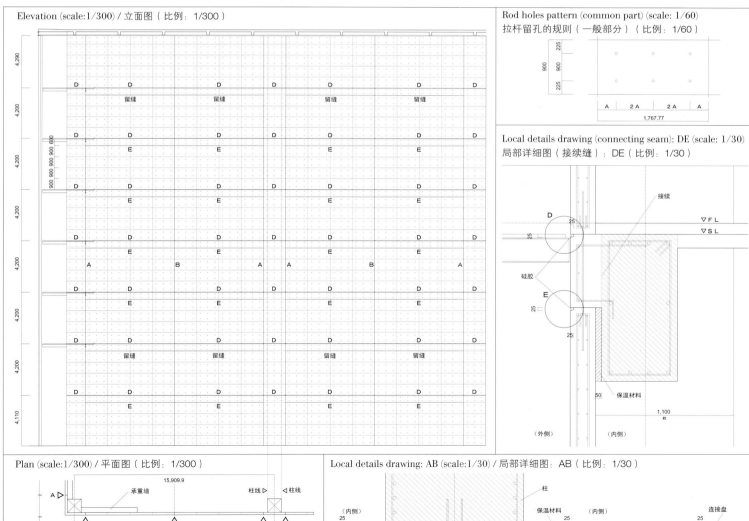

# Curtain Wall
幕墙

The theatre wall evokes the image of grey concrete wrapped in a fine scarf, imbuing the architecture with a sense of natural beauty. The space can also hold a diverse range of activities and performances. More importantly, the curtain wall is applied with respiration-type double-layer curtain wall, allowing for significant energy savings, emissions reduction, sound insulation, and noise reduction.

大剧院的幕墙宛如一层轻纱包裹着青灰色的混凝土墙体，让建筑若隐若现于自然天地之间，其间形成的空间也为大剧院的演出、参观等功能流线提供了多样的可能性，更为重要的是整个幕墙采用了双层呼吸式幕墙技术，为大剧院的节能减排、隔声降噪起到了至关重要的作用。

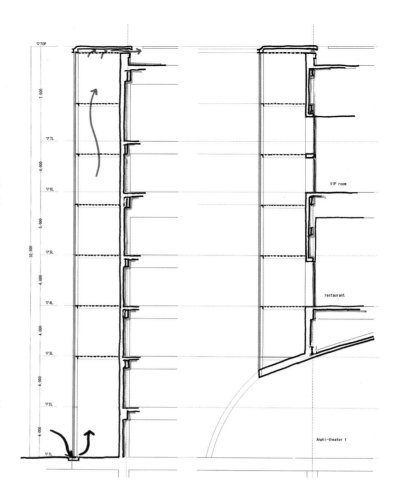

PP158-159: Installing the curtain walls
P160: Top, Study of the curtain wall structures; Bottom, Installation of the curtain walls at the Grand Theatre's northern side
P162: Ventilation layers separate the respiration-type double-layer curtain wall, glass curtain wall, and fair-faced concrete wall

158-159 页：幕墙安装施工现场
160 页：上，幕墙结构研究草图；下，大剧院北侧立面幕墙安装施工现场
162 页：呼吸式幕墙、玻璃幕墙和清水混凝土墙体之间的通风换气层

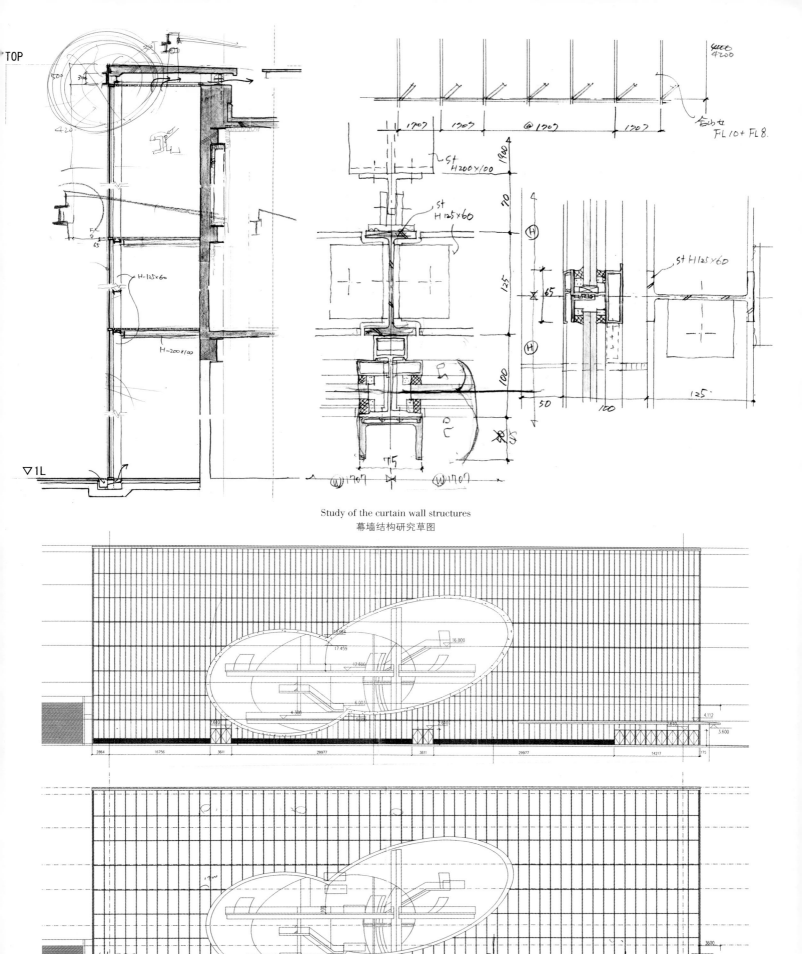

Study of the curtain wall structures
幕墙结构研究草图

Curtain wall façade segmentation comparison
幕墙立面分割对比图

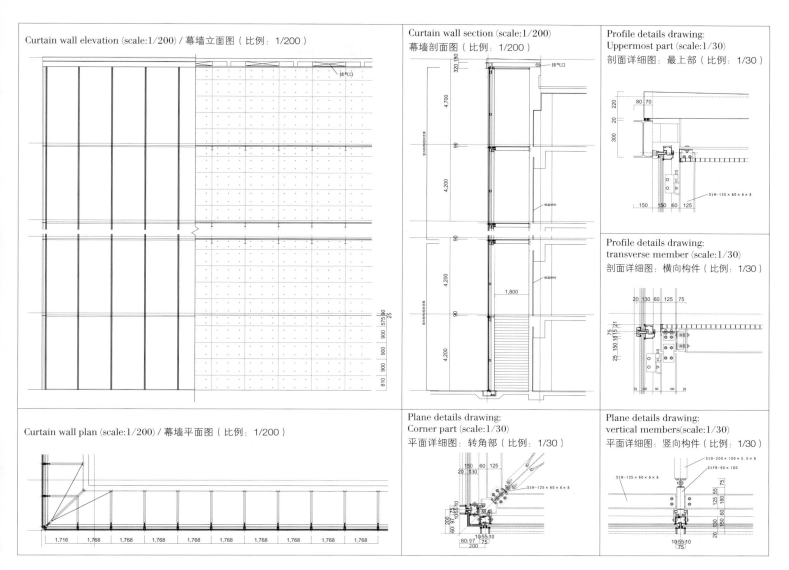

The theatre's curtain wall system is quite complex, with a total of twelve subsystems covering nearly an area of 30,000m². The main feature of the respiration-type double-layer curtain wall is the installation of ventilation layers between the glass curtain wall and the fair-faced concrete wall, with a mobile shuttering device installed at the top and at the foot of the curtain wall to control the device. The ventilation in these layers can help reduce the temperature of the fair-faced concrete wall to room temperature, and reduce the temperature difference between the walls. This design allows for additional energy savings of 42%-52% during heating and 38%-40% during cooling, when compared with traditional curtain walls. During design, simulations of the glass curtain wall were conducted to ensure minimal light pollution in the building.

大剧院的幕墙体系相当复杂，共有12种不同的幕墙系统，面积近3万 m²。其呼吸式幕墙的最大特点就是在玻璃幕墙和清水混凝土墙体之间形成一个通风换气层，并在幕墙的顶部和脚部设置活动百叶装置来控制其开合。通过换气层中空气的流通或循环的作用，使清水混凝土墙体接近室内温度、减小温差。因而它比传统的幕墙采暖时节约能源 42%~52%；制冷时节约能源 38%~40%。另外通过玻璃幕墙反射光模拟分析设计将建筑的光污染降至最低。

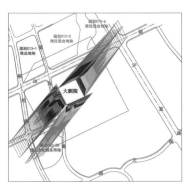

The coverage of the reflection lights on vernal / autumn equinox
春秋分反射光线影响范围

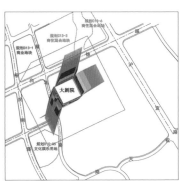

The coverage of the reflection lights on the summer solstice
夏至日反射光线影响范围

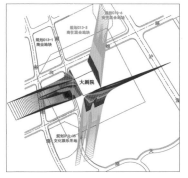

The coverage of the reflection lights on the winter solstice
冬至日反射光线影响范围

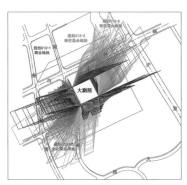

The coverage of the reflection lights for the whole year
全年反射光线影响范围

# Cylinder Nodes
## 圆筒节点

The design of the theatre intersects and combines three-dimensional cylindrical forms to create a space that is pure and powerful. The crisscrossing cylindrical forms generate large complex curves, which are the theatre's most lively architectural elements.

　　在大剧院的空间设计上，采用了"圆筒"立体穿插、组合的手法，形成了纯粹而又充满力量的空间。一个个功能圆筒所形成的相互交错咬合的大尺度复杂曲线，成为大剧院重要的公共空间中最具生命力的建筑表现。

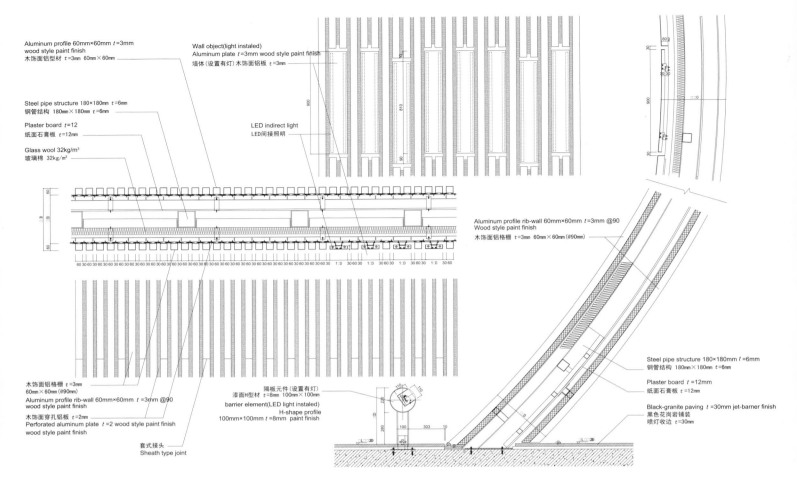

Rib-wall details (scale:1/30) / 墙面百叶详细图（比例：1/30）

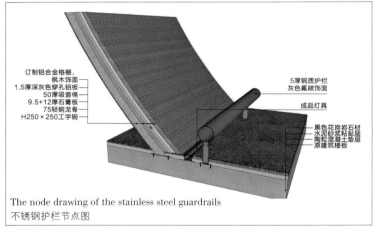

The node drawing of the stainless steel guardrails
不锈钢护栏节点图

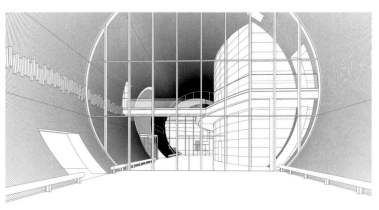

The wooden grille surface is uniformly applied onto the cylinders to maintain the theatre's purity and integrity. The wooden grille at the entrance hall is arranged horizontally, while those for the foyer and the indoor-outdoor platform are vertically aligned. The continuity of the space makes the nodes at the spatial junctions very complex. Using 3D modeling, a ridge is designed at the grille junctions, making the original shape much simpler and clearer. However, as the curves of the ridges are always complex in shape and very large in scale, they could only be perfected via an onsite BIM system, which perfected the spatial positioning of each curve after a repeated series of checks and adjustments.

为了让剧院内外的空间保持纯静的特质以及整体性，圆筒统一采用了木格栅装饰面。入口大厅的木格栅为横向设置，而观众候场空间以及半室外平台空间则采用了竖向排列。由于空间的连续性，所以空间交合部位的节点异常复杂。通过三维形体的确认，设计最终确定在格栅交接部位设置一条棱线收边，让原来复杂的形体变得简洁清晰。但是这些收口曲棱线部件不但形体复杂而且尺度都非常大，最终通过现场采用BIM系统进行空间精确定位和反复校对调整后得以最终完美实现。

PP164-165: Construction site of the cylindrical wall surfaces
PP167: Top, Construction site of the foyer; Bottom, Intersection of the respective cylindrical spaces of the entrance and foyer

164-165 页：圆筒墙体饰面施工现场
167 页：上，观众休息厅施工现场；下，入口大厅圆筒空间与休息厅圆筒空间相交处施工研究现场

P168: Cylindrical space with vertically-aligned wood grille surface
P169: Wood grille installation site

168 页：竖向排列木格栅圆筒空间
169 页：木格栅安装施工现场

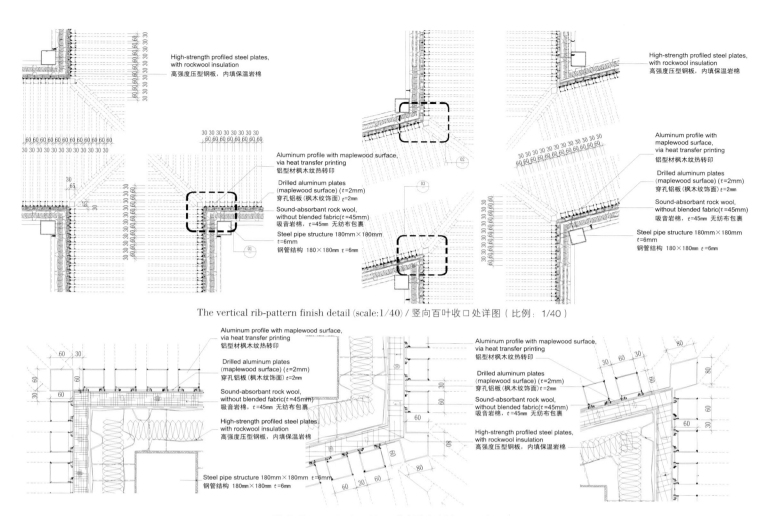

The vertical rib-pattern finish detail (scale:1/40) / 竖向百叶收口处详图（比例：1/40）

Node drawing (scale:1/60) / 节点图（比例：1/60）

# Interior Decoration of the Auditorium
观众厅室内装饰

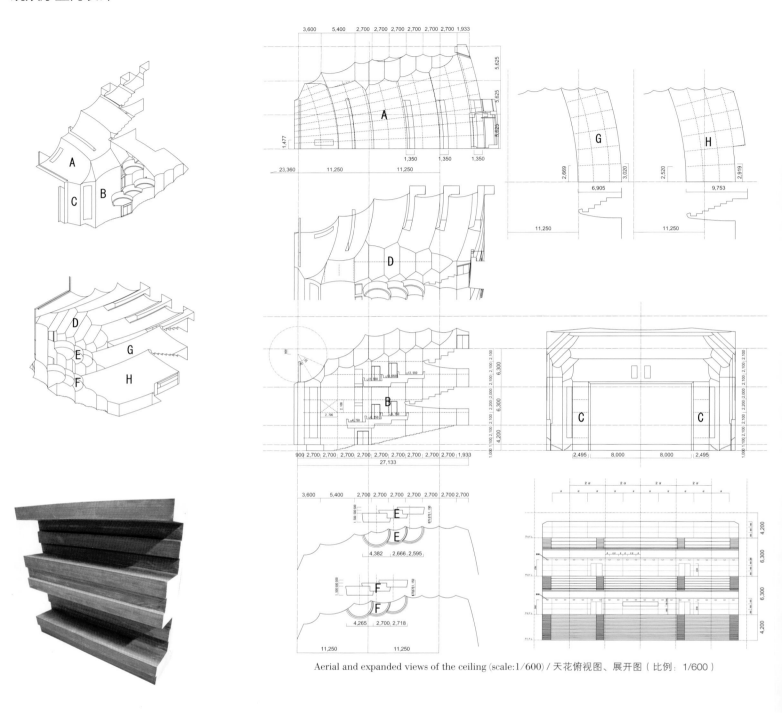

Aerial and expanded views of the ceiling (scale:1/600) / 天花俯视图、展开图（比例：1/600）

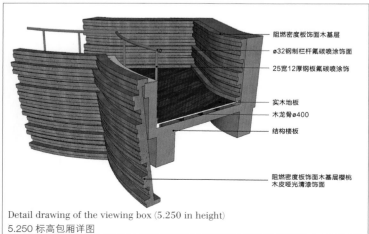

Detail drawing of the viewing box (5.250 in height)
5.250 标高包厢详图

The interior decoration of the auditorium has to, by definition, attain high visual and acoustic standards, while satisfying the strict fire safety regulations. In order to present the design concept of "delicate instruments", a series of concave and convex curves is applied to the wood walls. Such curvatures can improve the effect of reflected sound, provide the proper reception angles for listening, and add some lively elements to the serious setting of the auditorium.

观众厅的室内装饰不仅要求空间的高品质，更加需要满足视线、消防、声学等功能性的要求。为了呈现"精致乐器"这一设计理念，大剧院观众厅的室内空间采用了整体曲线凹凸木饰面的设计。这些翻卷的曲面造型能够提高反射声的效果，并为耳光（舞台灯光）提供适合的照射角度，同时也为比较严肃紧张的观众厅增添了一些活泼的氛围。

P171: Detail of the laminated pine surfaces
P172: The concave and convex curves of the auditorium's decorative wood surface

171页：松木集成材饰面细部
172页：观众厅曲线凹凸木饰面整体效果

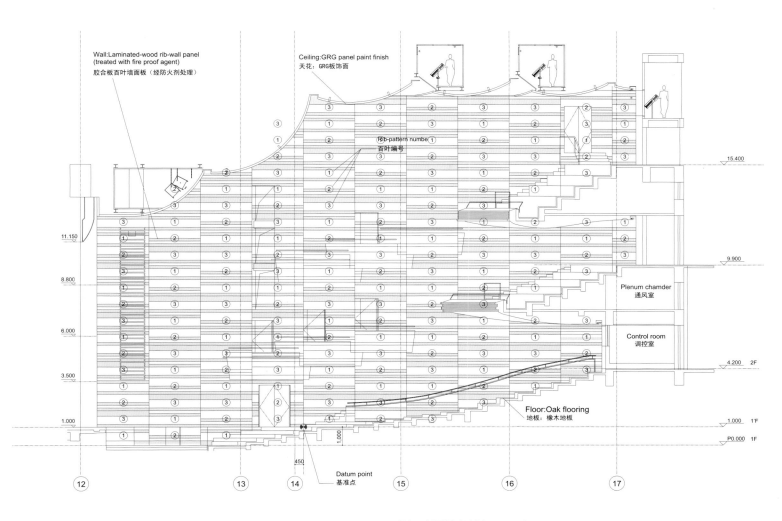

Auditorium section (scale:1/200) / 观众厅剖面图（比例：1/200）

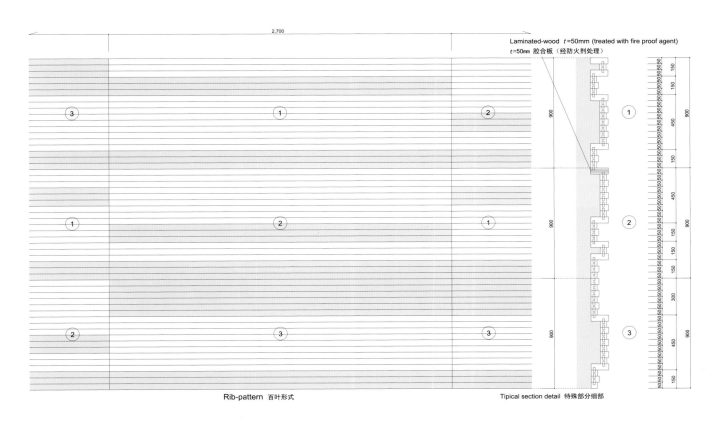

Rib-pattern 百叶形式　　　　　　　Tipical section detail 特殊部分细部

Auditorium rib-wall details (scale:1/30) / 观众厅百叶详细图（比例：1/30）

# Lighting
照明

The lighting of the theatre is simple, highlighting the nobility and elegance of the fair-faced concrete venue. Overlooking the lake, the vision extends to the horizon, and light and shadow are intertwined in graceful dance, while a cool breeze sets off ripples in the water. Voilà, a marriage of nature, light, and artifice. This interplay of the elements is reflected in the dialogue between the theatre and the water, where reality mingles with illusion, epitomising the concept of "moonlight in water". People do not merely walk on the lakeside passage: they walk along the "Milky Way", illuminated by the reflection of starlight in water. The breathing of light and the dance of shadows make the building come alive, transforming it into a living organism.

大剧院的泛光照明以简洁为原则，凸显清水建筑高贵典雅的气质。从湖岸远眺，视觉随着光延伸至天际，光和影婆娑流动、清风水面涟漪，让"光"与"建筑"、"自然"美妙地融合在一起，充分考虑"建筑"与"水"的虚实，掩映出"月光如水"的意境。人们漫步于休闲步道上，在水体星光的映衬下，如同"漫步银河"；光的呼吸、影的流转，让建筑变得更加鲜活，仿佛一个有生命的有机体。

Prelude—The Awakening: A beautiful interplay of space and light in the theatre's kaleidoscope
序曲——觉醒：安静而温暖的色调，万花筒般迷人的空间和倒影

Phase One—Listening to waves: Silvery light twinkling in the night
初章——听涛：光影追逐而下，如银辉洒落

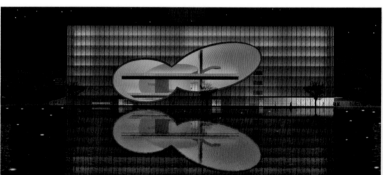

Phase Two—Foggy Forest: Ode to nature
中章——迷雾森林：水与自然的赞歌

Crescendo—Lunar Dream: Poem of light and shadow
高潮——月梦：聆听光影之诗

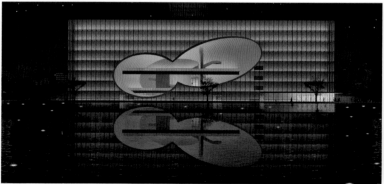

Epilogue—Starry Ocean: Light of the Future in all its elegance
终章——星海：优雅的未来之光

The normal effect of lighting
实景灯光效果

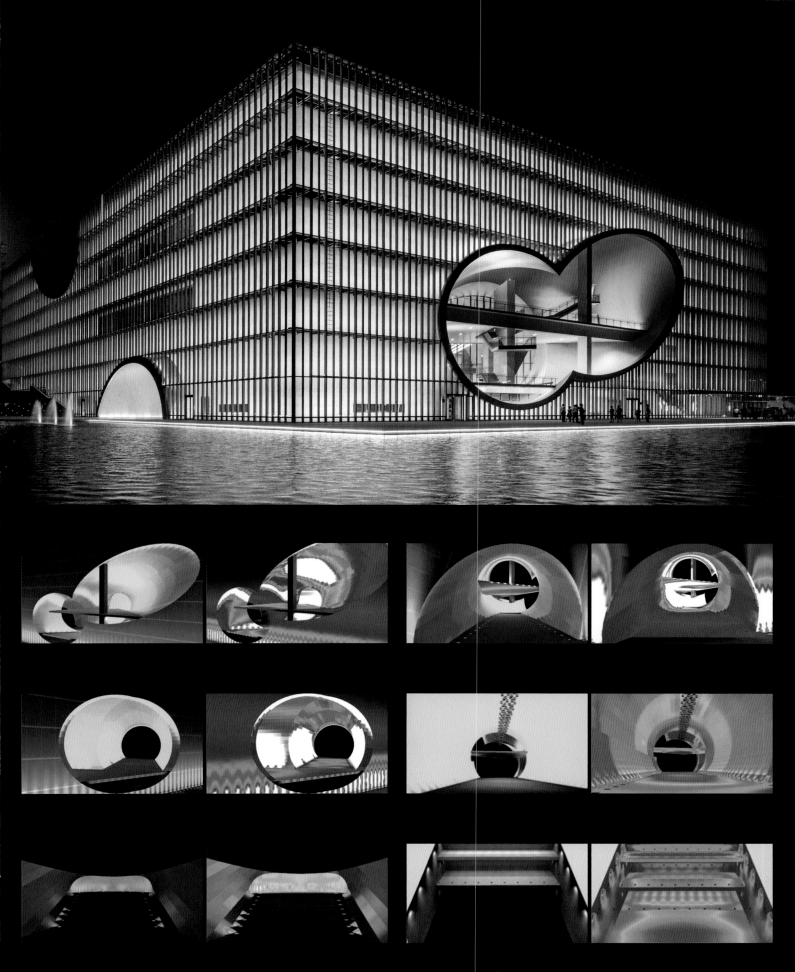

Local illumination sketch and lighting simulation rendering effect in key areas
重点区域灯具局部照明示意图与照明模拟效果渲染图

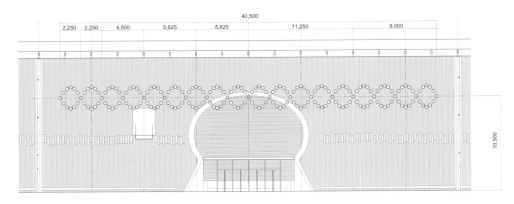

Expanded view of the lighting (scale:1/400) / 照明展开图（比例：1/400）

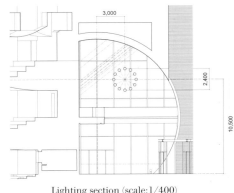

Lighting section (scale:1/400)
照明剖面图（比例：1/400）

Aside from the basic lighting fixtures, most of the building is illuminated by indirect lighting, continuing the space's theme of purity and simplicity. In particular, the double helical drop-lights which is over 40m long hanging in the cylindrical space of the foyer have become the space's crowning achievement.

建筑内部的照明除功能性需要之外，基本采用间接照明，最大限度地保证了建筑空间的纯净和简洁。值得一提的是作为观众厅前厅室内圆筒空间的视觉焦点，一组40多米长的双螺旋形设计吊灯成为点睛之笔。

PP176-177: Over 40m long, these double helical lights illuminate the foyer
176-177 页：休息厅室内 40 多米长的双螺旋形吊灯

# Theatre Acoustics
剧场声学

In order to achieve international standards for theatre acoustics, Makoto Karasawa, a famous theatre acoustics expert in Japan, was invited as a consultant to the initial design phase. Three different models were designed, and acoustics tests were conducted. After satisfactory results were obtained, theatre acoustics expert ZHANG Kuisheng then executed the design. Constant supervision and verification by experts in China and abroad ensured that the theatre met international standards in the following: reverberation frequencies, decay times, transparency, clarity, intensity, lateral reflection, and intimacy.

为了实现国际化水准的剧院声学效果，在建筑设计的初始阶段，就邀请了日本著名的剧院声学专家唐泽诚担任剧院声学顾问，先后三次制作了不同比例的观众厅模型进行声学实验。在取得满意的结果后，再由国内剧院声学专家章奎生进行专业的声学设计。通过国内外专家反复的验证和层层推敲，建成后的混响时间、早期衰减时间、明晰度、清晰度、强度、侧向反射声、亲切感等各项指标都取得了令人满意的效果。

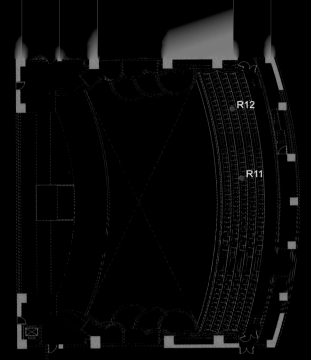

Layout of measuring points for third-floor seating
三层楼座测点布置图

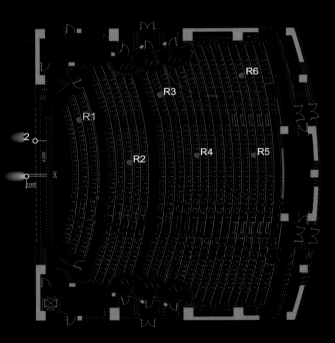

Layout of measuring points for first-floor seating
一层楼座测点布置图

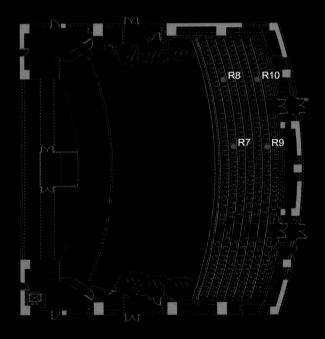

Layout of measuring points for second-floor seating
二层楼座测点布置图

## Opera 歌剧演出模式

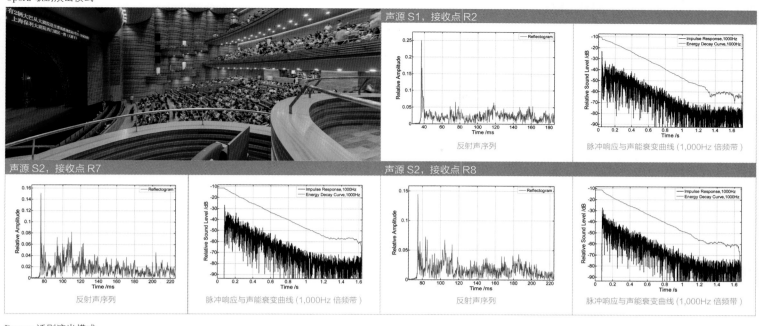

## Drama 话剧演出模式

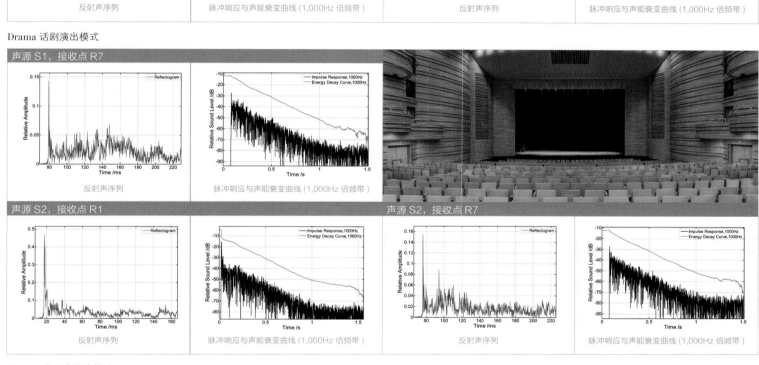

## Concert 音乐会演出模式

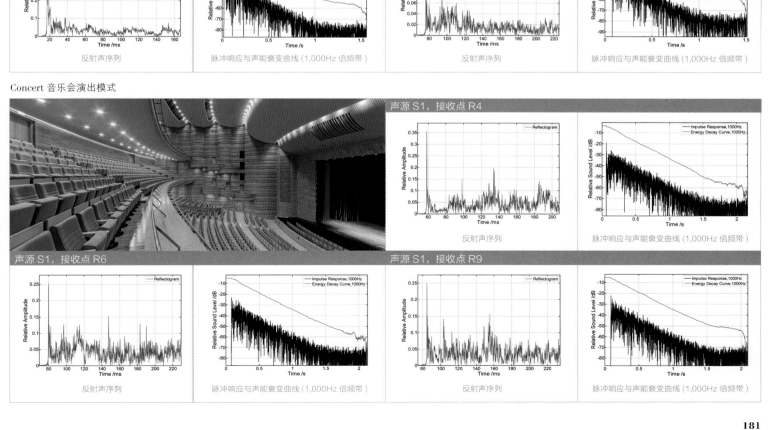

# The Theatre and Me
剧院·缘

ZHAO Guo-ang  赵国昂

Early in the summer of 2005, when the Dongguan Yulan Theatre was almost completed, I was asked to come to Shanghai from Guangdong. For the past five years, it had an honor to be involved in the construction of the Beijing Poly Theatre, the Dongguan Yulan Theatre, the Wuhan Qintai Grand Theatre, the Shenzhen Poly Theatre, and the Guangzhou Opera House. To me, theatre is like a stranger who became my old friend. When I left Guangdong, some of my colleagues ridiculed me, prompting me to quit the field of theatre architecture. Seeing that my work in Shanghai was to construct commercial buildings, I thought the relationship with my old friend had ended.

However, everything in the world is intangible. Jiading New City planned a series of cultural buildings. The smooth progress made during the early stages pushed forward the discussion of building a theatre in Jiading District. Poly Property focused on cultural real estate and was high recognised by the Jiading District Government. Thanks to the cooperation between the Poly Property Group and the government, I got another chance to build a theatre. I therefore met the Japanese architect guru, Tadao Ando.

Mr. Ando is called a magician of space. He is good at using fair-faced concrete to create simple forms. He focuses on getting architecture to harmonize with nature, mixing wind, water, light, and buildings with perfection. After the ellipsoid of the National Centre for the Performing Arts by Paul Andrew, the fishes of Wenzhou Grand Theatre by Carlos Ott, the smooth rocks of the Guangzhou Opera House by Zaha Hadid, the long-sleeve dance of the Zhujiang Architecture Design Institute and the Wuhan Qintai Grand Theatre, what will Tadao Ando's Shanghai Poly Grand Theatre be like? We fully expected a new friend to show up.

At first sight of the model of the Shanghai Poly Grand Theatre, I admired the district Government and the company executives for choosing Mr. Ando as the architect. Mr. Ando's design is truly unique. The model was a simple rectangular box, 100m square and 34m in height. The surface of the cube were the respiration-type glass curtain walls, making the box looking a crystal veil covering the fair-faced concrete. The cylinders cutting into the cube made it look like a cubical Taihu Stone. When you observe carefully, it looked like the cylinders were dancing in the box; it didn't matter if they were arranged vertically, horizontally, and diagonally. It didn't matter if they were intersecting one another, barely brushing one another, or cutting into the façade. The dancing forms defined the entrance hall, foyer, the traffic flow, the terrace amphitheatre and various other spaces. It is amazing to witness the perfect combination of functionality with unusual but wonderful thinking.

Outstanding design meant difficult construction. The first difficult parts were the 36,500m² fair-faced concrete. There were four test models, a specially-trained construction team, and over 10,000m³ of custom-made concrete. We had to choose the appropriate moulds and devise ways to protect the finished product. In an era defined by a lack of professional laborers, people only thought of speed instead of quality. That was why it was difficult. Secondly, Mr. Ando used crisscrossing cylinders as the theme of the design to make very complicated curves. These curves were the most important components in the architecture. They were not only complex shapes, but were also of long length. Some were more than 30m long. Therefore, the components' aluminum tubes could not be processed in the factory, meaning that the processing and adjustments were made onsite by the construction team. The decorative surface of the rectangular aluminum tubes occupied an area of 10,560m². The total length of the material reached 117,321.6m, which meant that if we connected the material one by one, it would cover the distance from Shanghai to Suzhou. Moreover, we were building the first solid wood auditorium in China. The area and number of decorative pine surfaces were amazing. The main walls, the fences, and the ground surface totaled nearly 3,300m². The number of units of the material reached 50,000.

On the 5th of September, the day of the first rehearsal at the Shanghai Poly Grand Theatre, Jiading District's huge treasure box opened itself to the world, its graceful beauty reflected in the surface on Yuanxiang Lake. On one side, one could sense its motionless tranquility; on the other side one could feel its gentle ripples on the lake. When I saw this combination of dynamic and static images, I thought of the entire theatre project, from the initial proposal to the confirmation of the mode of collaboration, from design to construction, from installation to debugging. Many people had worked hard on this site, with non-stop planning, great expectations, constant hesitation, and labouring under the burning sun. Suddenly, I was moved by the experience.

A few days ago, I went to Japan to study PC architecture. I was planning to go to Osaka on the 18th of September to meet Mr. Ando and view some of his projects. However, because of an urgent matter in the office on the 17th, I flew back to China. It was a pity that I could not see the architect guru's projects. Anyway, it is the unexpected nature of our opportunities, whether they are seized or missed, that makes life beautiful. As long as there is some chemistry between the two of us, we will meet again at another place.

(ZHAO Guo-ang  Poly Property Group (Shanghai) Investments Co., Limited CEO)

  2005年初夏，在东莞大剧院接近完工之际，我奉调自粤入沪。此前五年左右的时间，有幸参与北京保利剧院改造以及东莞玉兰大剧院、武汉琴台大剧院、深圳保利城市剧院、广州歌剧院等项目建设。剧院于我，由陌生人日渐成为老朋友。离粤时，几位业内同仁以调侃的口吻同我告别，说我是退出剧院这个江湖了。鉴于这次赴沪的任务是建设商业写字楼综合体，我也以为我与老朋友的缘分自此已尽。

  然而，世上的事情竟是如此地不可捉摸。嘉定新城规划了一系列文化建筑，而前期开发的顺利推进，使得在嘉定建一座歌剧院被提上议事日程。保利置业是专注文化地产的品牌，也得到嘉定区政府的高度认可。保利置业集团和政府的合作，把建造剧院的机会又一次推到我的面前，我也因此得以与日本建筑大师安藤忠雄先生邂逅。

  安藤先生号称空间魔术师，擅用混凝土塑造简洁形体，一贯重视建筑与自然的和谐相处，善于将风、水、光与建筑完美融合。在安藤鲁国家大剧院的椭球、卡罗斯温州大剧院的鱼、扎哈广州歌剧院的圆润砺石、珠江设计院武汉琴台大剧院的水袖长舞之后，由安藤先生担纲设计的上海保利大剧院会是一个什么样子呢？我们在忐忑中对这个新朋友的亮相充满期待。

  第一眼见到上海保利大剧院的模型，不禁感佩区政府和公司领导选择安藤先生的英明。安藤先生的设计的确与众不同，它是一个简单到极致的方盒子，底面边长各100m、高度34m。方盒子的外皮是呼吸式玻璃幕墙，恰似给混凝土披了一层晶莹剔透的轻纱。方盒子中穿插了几个圆筒，乍看上去，好似一个被削成正方体的太湖石。仔细观察，剧院的空间仿佛是几个圆筒在正方体内舞蹈，或站立或平躺或斜倚，或两两相贯，或为平面所截，或与立体轻擦。舞蹈的定格形成了剧院入口大厅、观众休息厅、交通动线、半室外剧院、屋顶舞台等多样性的功能空间。实体功能与奇思妙想如此完美地结合，令人惊叹。

  设计的出挑，对施工来说却是挑战。工程的难点，首先是3.65万m²的清水混凝土、四次样板试验、专门培训的建筑工人队伍、超过10,000m³的定制商品混凝土，还有模板的选用和成品保护，在建筑业缺乏产业工人、追求速度甚于质量的时代，难度可想而知。其次，安藤先生采用了圆筒相互交错咬合的设计形式，由此生成了非常复杂的交合曲线。而这些交合曲线是建筑中最为重要的表现构件，不但形体复杂而且长度都非常长，有些甚至达到30多米。所以这些构件的铝方管无法在工厂进行加工制作，只能在实地根据测量的数据进行加工和调整。铝方管装饰面积约10,560m²，使用材料总长度达到117,321.6m，也就是说单根材料相连，可以从上海连到苏州。还有，我们塑造了国内第一个实木观众厅，所用松木集成材的装饰面积和数量也是非常惊人的，主墙、八字围栏、地面完成面的面积近3,300m²，集成材和地板板材的数量达到5万多件之多。

  9月5日，当这座嘉定的巨型宝盒——上海保利大剧院，第一次试演的时候，打开泛光后的建筑体倒映在远香湖中，一边静静地伫立，另一边随微波荡漾。看着这两幅动静交融的图案，想想剧院项目从最初被提议到合作方式的确定，从设计到施工，从安装到调试，多少领导和同事的心血凝聚在这里，多少次的构划与憧憬，多少次踌躇与纠结，多少个烈日下的赶工，多少个不眠之夜的思索……突然，心中有一种说不出的感动。

  前几天又一次去日本考察PC建筑，本来计划在9月18日去大阪，拜会安藤先生并参观安藤先生设计的几个作品，却因公司有急事于17日匆匆回国，错过了再次欣赏大师作品的机会。遗憾之余，倏然冒出一个念头，不期而遇和失之交臂，都是人生的一种美丽，只要有缘，错过的一定会在另一个地方再相遇。

[ **赵国昂**　保利置业集团（上海）投资有限公司　董事长 ]

P182: The view of Shanghai Poly Grand Theatre
P183: The rich layering of the foyer

182 页：上海保利大剧院外观
183 页：层次丰富的休息厅空间

PP184-185: The foundation under construction
PP186-187: The cylindrical opening under construction

184-185 页：地基工程
186-187 页：圆筒开口施工现场

PP188-189: The construction team was working hard to meet the deadline
PP190-191: The entrance hall appeared spectacular even before the installation of the wood grille surfaces

188-189 页：建筑施工人员努力赶工中
190-191 页：未安装木格栅饰面之前的入口大厅已初显它的气魄

PP192-193: The construction site of the entrance hall
PP194-195: The cylindrical space under construction
192-193 页：入口大厅施工现场
194-195 页：圆筒空间施工现场

PP196-197: The complexity of the cylindrical space's curving walls required great concentration and care from the workers
PP198-205: Installation of the curving wood grille surfaces
PP206-207: Installation of stage machinery and equipment

196-197页：由于圆筒弧线较难施工，施工人员丝毫不敢大意
198-205页：圆筒弧线木格栅饰面安装
206-207页：舞台机械设备安装

CHAPTER 3 | 第 3 章

# Technological Realization
# 技术实现

**Plan Drawings**
方案的图纸实现

**Fair-faced Concrete**
清水混凝土实现

**Interior Decoration**
室内装饰实现

**Stage Mechanics**
舞台机械实现

# Pearls Strewn in Concrete and Soil
散落在营造中的珠玑

SUN Jian 孙健

A building represented in various technical forms during the construction process will be diluted slowly with time. However, such seemingly isolated elements will be revived and refreshed if strung together in a line of thought.

On September 30, 2014, the symphony concert entitled *Poly Property: The Evening Gala* at Jiading New City, played by the German Radio Philharmonic Orchestra, and directed by Chung Myung-whun, made its debut at the Opera Hall of the Shanghai Poly Grand Theatre (SHPGT). It also marks a new beginning of her life journey.

After five years' construction, the "waterscape theatre", a rich, elegant space comprising of simple geometries and epitomizing Mr. Ando's design concept of the "seamless merger of artifice and nature", finally rose up beside Yuanxiang Lake in Jiading, Shanghai, unveiling a kaleidoscope that demonstrates the "genuine harmony between nature, people and culture". Anyone the theatre will be attracted by the orderly variable spaces of the "cylinder" and touched by an atmosphere in which lights and shadows are interlaced. The anecdotes I will talk about are exactly the pearls strewn in during the construction.

**Tone quality: the soul of architecture for the performing arts**
Excellent tone quality is the soul of theatre construction. Architectural acoustics is the key to the entrance of the "soul". However, excellent tone quality is hardly controllable in such an unpredictable construction process.

To improve the acoustic quality of the theatre, designing a shape in line with acoustic rules is critical. For this purpose, we worked together with Tadao Ando Architect & Associates, the Tongji Architecural Design (Group) Co.,Ltd and ZHANG Kuisheng Acoustic Design and Research Studio to research on the relations among the plane, number of seats, capacity and functional positioning of the auditorium to establish an integrated system where plane, volume and acoustics support each other organically. To make the architectural acoustics more accurate, we further conducted a 1/30 model test with Tadao Ando Architect & Associates and Makoto Karasawa Architectural & Acoustic Design to verify our prior research achievements after the initial draft of the computer acoustic simulation report was completed. Although the two acoustic research institutes worked independently, their research results were checked and verified with each other for several times thereafter. With the differences in their "research results" gradually smoothened away, the plane and volume of the construction were basically determined. In this round of design, a gable was added in the rear rows at the cost of adjusting the capacity of the auditorium and sacrificing nearly 5% of the auditorium seating on the first floor to enhance the early-stage echoes at the middle and rear zones. These measures were further substantiated by test data in the acoustic measurement report after conducted after the project was completed, and they laid a solid foundation for the rapid progress and excellent tone quality of the project.

In the detailing of the interior decoration design, ZHANG Kuisheng Acoustic Design and Research Studio conducted a 1/10 model production and a computer acoustics simulation to verify the early-stage design achievements in a more profound and accurate manner to ensure excellent tone quality. Numerous adjustments and computations benefited the interior decoration design.

No pain, no gain. The nine major indexes in the acoustic measurement report further convinced the contractor that the project was "the best" in Shanghai. Conductor Chung Myung-whun, performer FU Cong & ZHANG Haochen and senior tuners in Shanghai all praised the "music quality" of the Theatre.

**Saving Energy: the shouldering of social responsibilities**
A theatre is a large public construction and therefore has to conform to the *Design Standard for Energy Efficiency of Public Buildings*. To save energy, research was conducted on various energy-saving technologies with reference to experiences of previous projects and the initial investment budget, such as natural lighting (for the entrance hall, foyer, offices, and makeup zone) and ventilation. Below is a summary of all the energy-saving technologies we looked at, and the details of their implementation in the project: Breathable walls: Openable blinds were designed to be open in the summer for cooling and closed in the winter to preserve heat; Air-conditioning: The "cold storage" technology is adopted as the cooling source; Water pump fans: Variable-speed regulation is adopted; Layered temperature control in the auditorium: CFD-aided simulation technology was used during the early design stage; Air monitoring: Sensors are mounted on the walls of the auditorium for interior air data collection, for the monitoring of the status of air feeding/exhaust devices, and to make automatic adjustments of fresh air input throughout the space. Garage sensors: When the concentration of carbon monoxide reaches a critical level in the garage, the alarm will sound and the system to activate the exhaust device; Rainwater recycling: Rainwater on the rooftop is recycled, raising the percentage of unconventional water sources in structure to 61.53%; Rooftop greening: The vegetation's transpiration effects enhances heat insulation, reducing the load on the air-conditioning system; These energy-saving technologies are applied to improve the internal environment and reduce energy consumption to the greatest extent possible, while remaining economical and efficient. Thereafter, the project progressed toward "maturity and efficiency". By chance, we started to analyse our green architecture data, as recommended by the Guangzhou Institute of Energy Conservation and the Chinese Academy of Sciences. We were surpassed to find that the application of such "mature" technologies enabled us to conform to the two-star green architecture standards without adding any equipment or modifying the architectural plane. So we applied for the two-star green architecture certification. I believe that this is the return to our "sense of responsibility". The application is currently in the final stage of expert review.

**Fair-faced concrete: the Zen of the East**
"The calm after the storm": the highest praise of fair-faced concrete, the "Zen of the East". It is difficult to realize the fair-faced concrete walls with only two parallel lines on the design drawing.

In the early preparatory stages of fair-faced concrete construction, the team had a limited understanding of the process and consequently underestimated the difficulties encountered. To stimulate the young team's morale, we formed a three-party QC technical team with the supervisor and the contractor and started multiple rounds of discussions and tests on seven primary aspects of quality control, i.e. concrete, reinforcing bar, molding board, casting and tamping, curing and demolding, formation of doors, windows and equipment as well as surface protection. With

建筑在其建造过程中会以各种的技术形态呈现,并随时间的推移慢慢褪色。但在我看来,这看似孤立的片段,若用一条思想的线将它们串起,放眼望去,眼前的景象又会鲜活生动起来。

2014年9月30日,在上海保利大剧院歌剧厅,由郑明勋指挥的德国广播爱乐乐团《保利置业·嘉定新城之夜》交响音乐会拉开了开业首演的序幕,同时开启了她的生命旅程。

历时五年的建设,一座承载着安藤先生"建筑与自然环境无间融合"的理念,在由简单几何学生成的多变华丽空间里,实现了"自然与人、与文化的碰撞",并呈现出"万花筒一般"丰富景象的"水景剧院"已矗立在上海嘉定的远香湖畔。

走入今天的剧院,会被"圆筒"有序多变的空间所吸引,也会被光影交错的氛围所震撼与感动,而我要说的是建设中的轶事,也就是散落在营造中的珠玑。

**音质——演艺建筑的灵魂**

优良的音质是剧院建筑的灵魂,建筑声学就是打开这扇"灵魂"之门的钥匙,然而优良的音质有着可遇不可求的偶然性,这也是此类建筑难以把控的地方。

为了提高剧院的声音品质和实现项目的快速推进,符合声学规律的形体研究是关键。为此,我们在项目初期就会同安藤忠雄建筑设计研究所、同济大学建筑设计研究院、章奎生建筑声学研究所对观众厅的平面、座椅数、容积以及功能定位等相互关系开展研究,旨在建立平面、体积与功能等与声学互为支撑有机的整体关系。为了让建筑声学的成果更精确,在声学所计算机声学模拟报告初稿完成后,我们会同安藤忠雄建筑设计研究所与唐泽诚建筑声学事务所又开展了1/30的缩尺模型测试研究工作,以便验证前期的工作成果。此后,虽然两家声学所均独立工作,但成果相互校核验证,经过多次校验和"成果"的"差异"趋同后,建筑平面和体积等也基本确定。在这轮设计中采用调整观众厅容积以及损失一楼观众厅近5%的座位为代价,新增了位于后排的八字墙,用于增强中后区的早期反射声。这些措施在竣工后的声学实测报告测试数据中得以验证,也为项目快速推进和建声优良的音质奠定了基础。

在内装设计深化施工图阶段,为确保观众厅的音质品质,结合前期的成果,章奎生声学设计研究所又开展了1/10的缩尺模型制作和测试以及计算机声学模拟工作,更深入和精确地验证前期的设计成果。无数次的调整和验算,为内装设计铺平了道路。

"宝剑锋从磨砺出,梅花香自苦寒来",竣工后声学实测报告的9项主要指标的结果给了建设者"上海第一,没有之一"的满意答案。指挥家郑明勋、演奏家傅聪、张昊辰以及在沪的资深调音师等都对剧院的"音质"啧啧称赞。

**节能——社会责任的担当**

剧院是大型公共建筑,满足《公共建筑节能设计标准》本是社会责任担当的一部分。为了更好实现节能目标,在建设初期结合已建工程的经验和本项目投资可能,开展了各项节能技术的研究,如自然采光与通风——在入口大厅、观众休息前厅、办公与化妆区等功能空间白天可采用自然光采光,过渡季节则可采用自然通风;可呼吸的幕墙——通过可启合的百叶窗,夏季开启百叶窗降低温度,冬季关闭百叶窗起保温作用;空调系统——空调冷源采用"冰蓄冷"技术,水泵风机采用变频调速控制方式,在深化施工阶段采用CFD辅助模拟技术指导,实现观众厅内温度分层控制;空气监测——观众厅利用装在墙上的传感器对室内空气进行数据采集,实时报警和检测进排风设备的工作状态,实现自动新风量调节;车库——当一氧化碳浓度达到报警值时,系统报警并相应开启排风;雨水回收利用——对屋顶优质雨水进行收集处理回用,非传统水源利用率达到61.53%;屋顶绿化——夏季通过植物的蒸腾作用增强屋顶的隔热功能,降低空调负荷。这些节能技术的应用目的是以最小且有效的投资,尽最大可能改善室内环境以及降低建筑能耗,之后项目也按既定"成熟、有效"的目标稳步推进。一个偶然的机会,在中科院节能所的推动下,我们开始绿色建筑资料的整理工作,结果惊奇地发现,这些"成熟"技术的应用可在不增加任何设备和不修改建筑平面的情况下,完全符合绿色二星建筑标准,这给了大家一个意想不到的惊喜,也开始了绿色二星认证的申报历程,我想这应该是"担当"给予的回报。目前剧院项目已完成申报工作,进入最后的专家评审阶段。

reference to the experiences of similar domestic and foreign projects and the results of four sample tests, after six months of hard work we finally established the *Manual for Management of Fair-faced Concrete of SHPGT* (Version 3), which included "analysis of technical difficulties, testing methods, practical principles, general process allocation, and personnel training". To ensure the quality of fair-faced concrete, we conducted "induction training" for our construction personnel according to features of this project, i.e. having the preliminary test results as the teaching materials and providing special trainings to the concrete workers, reinforcing bar workers, carpenters and E & M pre-burying workers. A certificate of excellence was issued when they passed the exam and only certified workers were allowed to start working. Moreover, we devised a flow chart summarizing fair-faced concrete project quality control for ensuring technical control over weak points in the process. These measures enabled the smooth construction of 36,500m$^2$ of concrete and served as a reference to the project's meticulous management. Surely, this only scratches the tip of the iceberg when it comes to the scope of technical difficulties encountered in the construction process.

## Aluminum alloy grille with wooden textures: the secret of a perfect solution

In the master's design, the interior wall surfaces of the cylinder should be shrouded by natural timbers. However, the "entrance hall" is a safety zone where the veneer materials must be Class A incombustible materials as a prerequisite for fire protection certification. But the existing synthetic wooden materials are Class B1 flame-retardant materials at the highest rating of fire resistance. For that reason, inflammable metal materials are the only solution. To achieve genuine effects of natural wooden textures, we attempted mixing metal and wooden materials together, but the effects were not satisfactory enough so we gave up the attempt. Metal + coating was adopted, hence leading to what it looks like today.

Such "solid timber" material has a special name on the worksite, i.e. "aluminum rectangular tube". Nobody knows the origin of it. Everyone was deeply impressed by the nature and shape of the material at first. It has a complicated full name, which is the combination of five main manufacturing techniques, i.e. customized molding, profiling, cold bending and forming, propylene powder spray coating and heat transfer of wooden textures. These different working procedures involve metal processing, forming and surface coating to be executed in different factories. Besides the 4m-long 60×60mm cambered "aluminum rectangular tube" material, we need other main materials for the veneer of the cylinder wall surfaces, i.e. the 3cm-high perforated aluminum panels coupled to the "aluminum rectangular tube", and the various types of "doors" on the cylinder and the intersecting lines between the cylinders. These "irregular plates" are the hard technical points unique to this project.

Besides difficulties in processing and manufacturing, installation also proved to be difficult. These seemingly isolated difficulties are interrelated. Before the construction of the floor structure, we already started research on the spatial relations between the BIM structure and cylinder keel. Through BIM modeling, we established the coordinates determining the position of the cylinder's central axis to the exterior curtain walls, ascertained the relations between the reservation and pre-burial points of the cylinder's keel structures and the curtain wall's legs, and provided the theoretical data for construction measurement and pegging. Surely, accurate construction is possible only through dynamic adjustments of actual errors.

## Solid-wood interior walls of the auditorium: past and present

The "solid wood" on the interior walls of the auditorium is also legendary. It used to be Class B1 keels of professional sports flooring which were not applicable as veneers. It is difficult because the wall surfaces of the auditorium needed to satisfy Class B1 flame retardance standards, relevant environment protection standards, the acoustic surface density requirements as well as overcome the materials' deformation. During construction, we conducted an experiment on "GRG + 3m wood-textured paper" and "Class B1 fireproof board + flame-retardant wooden veneer" but the test results were not satisfactory due to technical defects: Some veneer materials risked foaming; some techniques could not satisfy the requirements on 3D pasting requirements; some were too costly. When in a dilemma, we found the keel samples brought back from a factory two months ago having a transverse texture similar to that of oak, and we knew it was the right answer: its Class B1 flame retardance rating, synthetic solid-wood features presented a simple solution to the problem of thermal expansion. Meanwhile, its surface density could satisfy the acoustic requirements by adjusting its thickness. Most importantly, its building cost was really "affordable". Ideal material!

However, there was still a long way to go to turn the materials into products. Not only were there surface defects arising from processing and cutting, but the coating of curving surfaces on the assembly line had to be addressed. Not only did we have to satisfy the pro-forma inspection report on fire protection (original board thickness was 40mm, current thickness is 50mm), but we had to formulate the fire protection test report on field samples before we could use the materials in construction. It was also necessary to research the structure of connections to the walls and the resistance against insects. These technical problems were addressed; however, during detailed field design, the material processing specifications and quantities were highly different due to different installation plans. After some discussion, the factory promptly chose the processing plan with 1,300+ specifications, more than triple those of the original plan. Sincere pursuit of "construction quality". I was deeply moved. Now, it has become reality!

## Encouragement from the master

Mr. Ando attended one plan report and paid five personal visits to the worksite to direct the construction which really encouraged our team. During the correspondence with the master, Mr. Ando praised the orderly management of our worksite. Moreover, he and his peers were deeply impressed and surprised by the accomplishment rate.

On October 18, 2013, Mr. Ando expressed his opinions during an interview with *ESQUIRE*, "The world is preoccupied with the principles of economics and therefore it is difficult to finish a building that touches the heart of people. A construction needs to tell our soul and bring us hope... I visited the worksite of SHPGT this morning and I was deeply moved by the smooth progress of such a hard project. I know opposition from the neighboring residents and the difficulty in construction really puzzled the project owner and contractor. But their strong sense of honours has benefited them. I believe that it's unrelated with the principles of economics. The ardor and dedication of people will surely create an architectural heritage for generations".

## Thanksgiving

"Inherit civilization with construction, underscore harmony with wisdom and construct with a sense of responsibility for social service and in the role as a city developer and operator", as stated by Mr. XUE Ming, vice general manager of China Poly Group Corporation, during the foundation-laying ceremony of the project. This is also our guideline.

Everything starts from scratch, and so does this project. The completion of this theatre epitomizes the dedication and diligence of all contributors. Special thanks to the progress of the country and the development of our company because they provide each of us with an "opportunity." Special thanks to the government, our senior management and each member of our construction team for their support, assistance and participation in the project.

(SUN Jian  General manager/Chief engineer of Shanghai Poly Grand Theatre Project)

## 清水混凝土——东方禅学

"绚烂至极归于平淡",这是对有着"东方禅学"美称的清水混凝土的极高赞誉。要将设计平面图纸上仅是两条平行线的清水混凝土墙体变为现实,并不是一件简单的事。

记得在清水混凝土施工早期准备阶段,施工团队对清水混凝土施工的难度认识有限,对施工将遇到的困难也估计不足。为了让年轻的施工团队能真正接近临战状态,我们会同监理单位、施工单位成立了QC三方质量管理技术小组,开始了艰难的攻坚。从混凝土、钢筋、模板、浇捣、养护拆模、门窗与设备构造以及表面保护7个主要方面的质量控制点进行多轮研讨与试验,结合国内外同类项目考察和4段样板试验,历经6个多月,形成了包含"难点分析,试验方法,实施原则,总体部署,培训计划"等主要内容的第3版《上海保利大剧院清水混凝土管理手册》,至此基本完成了清水混凝土施工的前期准备。为了保证清水混凝土的施工质量,我们又开展了针对本工程特点的施工人员"岗前培训"活动,即以前期的试验成果为教材,对混凝土工、钢筋工、木工和机电预埋等工种的施工人员进行专项培训,考试合格后颁发专项合格证,实行"持证上岗"制度。此外我们归纳提炼了一张清水混凝土工程质量控制流程图,以便对质量难点进行分工艺控制。上述措施为顺利开展本工程3.65万$m^2$的清水混凝土施工奠定了基础,同时也为整个工程施工的精细化管理提供了可参照的样板。

当然,这些仅仅是施工技术实践的一个缩影,实际施工中的清水墙、楼梯等形式更为复杂,各式构件与结构或清水结构之间连接的施工方案、保障措施和遇到的困难等等,都比原先预想的要困难得多。

## 木纹铝合金格栅——难点的探密

在大师的设计里,圆筒内壁的表面应该是被自然木材包裹的。但是在消防审批中"入口大厅"为准安全区,区内的饰面材料必须为A级不燃材料,而现有的木质复合材料最高等级为B1级阻燃材料,为此只能从不燃的金属材料入手。期间为了实现真实的自然木纹效果,我们也尝试了将金属与木质材料相结合,但由于效果不理想,最终放弃了这个方向的努力,材料的试样朝着金属+涂层的方向发展,于是才有了今天的模样。

这种"实木"材料在工地上有个特别的名字:"铝方通"。也不知道是谁起的名称,何时开始就这样称呼了。但有一点,材料的性质和她的形状从她诞生开始就深深地印在大家的脑子里了。先说这种材料的全称——"开模热挤冷弯丙烯粉末喷涂木纹热转印铝型材",这是5个主要制造工艺的集成名称。它们的主要加工工艺是:定制开模、型材加工、冷弯成形、丙烯粉末喷涂、木纹热转印。这些不同工序,有的属于金属的加工和成形,有的是表面涂装,都要在不同的工厂才能完成。到此仅仅只是完成了长4m、截面为60mm×60mm的弧形"铝方通"材料,还有诸如构成筒壁饰面的主要材料——与"铝方通"联接的3cm高打孔铝板材,以及圆筒上各式的"门"和圆筒之间相撞形成的交线等,这些"异形的钣金"构件都是这个项目独有的技术难点。

除了加工制造难以外,安装也有着不同寻常的难度,看似孤立的困难其实是环环相扣的。因此早在地面结构施工前,我们已经开始着手开展

BIM结构与圆筒龙骨的空间关系研究,通过BIM模型建立了圆筒中心轴与外幕墙空间定位关系的坐标,同时为圆筒龙骨结构、幕墙牛腿等预留预埋建立了相对关系,也提供了施工测量放线的理论数据,当然在施工中通过对实际误差的动态调整,才能准确地施工。

## 观众厅实木内墙——前世与今身

观众厅内墙的"实木"采用也极富传奇性,起初她的"前身"是专业体育地板的B1级龙骨,但用于饰面似乎不太可能。由于观众厅的墙面既要达到阻燃B1级,又要环保,还要满足声学的面密度要求,同时还要克服材料的变形量,实属不易。在施工过程中,我们对"GRG加3m木纹纸"和"B1级防火板加阻燃木饰面"等材料进行了实验,发现效果都不理想,存在工艺缺陷。如有的饰面材料有起泡风险;有的工艺上无法满足立体粘贴要求;有的则造价不菲。正在一筹莫展时,我看见了办公桌上两个月前考察工厂带回的龙骨样品,眼前一亮。她有着和橡木近似的横纹观感,是B1级的阻燃材料,因为是复合实木特性,所以解决热胀冷缩的构造问题会相对简单,还有面密度通过厚度调整能满足声学要求,关键是造价也"亲民",这真是天赐的理想材料。

然而要把材料变成产品还有很长的路要走。从材料本身来看,不仅要解决材料表观由于加工切割形成的表面缺损的修补工艺,还要解决曲面成品在流水线上的涂装;不仅要符合消防形式检验报告(原来产品板厚是40mm,现为50mm),还要做现场抽检的消防检测报告后方可用于施工,同时还要研究与墙体的连接构造、防蛀防虫性能等问题。在技术问题一一落实后,现场深化又发现切割加工由于安装方案的不同,材料加工规格数量相差极大。通过沟通,为确保完成效果,厂方无怨无悔地选择了规格比原方案多3倍的1,300多种规格的加工方案。这是对"施工品质"的虔诚追求,让我深受感动。如今她终于"露脸"了!

## 大师的鼓励

建设期间安藤先生除一次参加方案汇报外,曾五次前来施工现场关心和指导建设,给予建设团队许多的关怀和鼓励。

在与大师往来的信件中,安藤先生对于我们将现场管理得如此井井有条而赞不绝口,同时该项目最终的施工完成度,让他和日本同行都惊讶不已,并深受感动。

2013年10月18日,安藤先生在接受ESQUIRE杂志采访时说:"在现今的世界中渗透了经济学原理,要完成一件能留在人们心中的建筑是非常困难的。建筑应该能叙述人们的心灵,给人们带来希望……今早我去了上海保利大剧院的工地,如此之难的项目能顺利推进,让我深受感动。我知道周边的反对和施工的难度给建设者们带来了很大困难,但业主团队和建设者们用荣誉感来跨越这重重困难。我相信这与经济学原理无关,人们心中的热情和执着,一定会创造出永留在人们心中的建筑。"

## 感恩

"以建筑传承文明,以智慧彰显和谐,以服务社会的责任感,和城市开发运营商的服务角色精心建设"是中国保利集团副总经理雪明先生在奠基典礼上的致辞,也是我们的行动指南。

合抱之木,生于毫末;九层之台,起于累土;千里之行,始于足下。剧院建筑的建成浓缩了全体建设者的执着与努力。

感恩国家的进步、企业的发展给予我们每个人的"机遇",感恩支持、帮助、参与项目建设的政府、集团、公司领导及同仁和所有建设团队的小伙伴们。

[孙健 上海保利大剧院项目总经理/总工程师]

P211: Installation of the curving wood grille surfaces
P213: Installation of the auditorium's decorative wood surface

211页:圆筒弧线木格栅饰面安装
213页:观众厅木饰面安装

# Plan Drawings
方案的图纸实现

CHEN Jianqiu 陈剑秋 / QI xin 戚鑫

Architecture is the art of space, an expression of the architects' thoughts and emotions, a messenger of plurality, depth, continuity, and dynamism. If architecture is frozen music, it is the countless drawings made by architects that record its movements.
We are honored and grateful for the chance to work with world-class architect Tadao Ando and Tadao Ando Architect & Associates through the Poly Grand Theatre project, witnessing firsthand how the master incorporates every ingenious idea into the architecture—a veritable labor of persistence. As an excellent domestic design institute, Tongji Architectural Design (Group) Co., Ltd also leveraged its expertise in theatre design, and project management to assist the architect in the creation of a world-class music hall for Shanghai and China.
The vehicle of communication among architects is the drawings, followed by words. In spite of a regional and linguistic differences, the drawings established a platform of exchange between Chinese and Japanese designers, becoming a "common language" that transcends geographical and cultural differences. From each sketch of the research phase to thousands of blueprints during the construction period, it is only five years, during which every detail is exhaustively examined, that a product satisfying all parties is created. Recalling fondly upon this design period, we remember Mr. Ando capturing everyone's imagination with a few lines when he proposed the prototype of the Shanghai Poly Grand Theatre in 2009. The infectious image still lingers: a rectangular form with a complex inner structure and pure surface, a cylindrical entrance endowed with a sense of ceremony, a vibrant lobby formed by crisscrossing cylindrical shapes, an amphitheatre by the water, and a terrace amphitheatre. Even now, the Theatre is presented as an original concept, and its degree of completion amazes people to this day.
On the one hand, it reflects the architects' comprehensive work during the initial design stage, and the high level of foresight and coordination required to overcome great technical difficulties during the construction period. On the other hand, it is the first time that anyone in China has ever constructed a fair-faced concrete building with a volume exceeding 36,500m². Its geometric complexity and large curving surfaces posed great difficulties during construction, making it a truly arduous journey to realize the architects' vision as presented in the drawings. However, Mr. Ando never wavered in his dedication to the original concept, transforming it into a building with clear outlines and a shining lustre. This emotional and dedicated approach to creation is a testament to his "spirit of craftsmanship," and it is this dedication that makes his work so infectious.
Our team has always been infected by such spirit during our cooperation with Tadao Ando Architect & Associates. In order to retain the moving power of the main theatre space, guarantee its efficiency as a high-quality performance space with complex technology, and incorporate the artistry of the visual lines into the living work, we had to maintain a rigorous work ethic throughout the entire design and construction processes.
To ensure an optimal visual experience at the Theatre, we drew upon

建筑是空间的艺术，它蕴涵着建筑师的思想和情感，它所传达的信息是多元的、多层次的、持续的和动态的。如果说建筑是凝固的音乐，那么记录这华美乐章的便是建筑师笔下那数不清的"图纸"。
很荣幸也很感恩通过保利大剧院这个平台有机会和世界级的建筑设计大师安藤忠雄及其建筑研究所来进行近距离的交流和合作，亲身经历大师是如何从精准巧妙的构思，一步一步坚持不懈，最终落实到每一个细节当中，同时也发挥同济大学建筑设计研究院作为优秀的本土设计单位在观演建筑设计上的优势和对大型复杂项目设计全过程的整合和把控能力，共同协助业主，为上海乃至中国呈现一个世界级的音乐殿堂。
建筑师之间交流的媒介首先是图纸，其次才是语言。虽然我们之间的合作有地域和语言上的不便，但是图纸为中日设计师之间的交流搭建了一个互通、互动、互信的平台，成为跨越地域和文化差异的"通用语言"。从方案深化阶段的每一张草图到施工图设计阶段上千张蓝图，中日双方的设计团队历时5年，孜孜不倦地不放过任何一个细节，才有了今天令社会各方都非常满意的作品呈现。
回顾这段令人难忘的设计历程，我们发现早在2009年底安藤先生第一次提出保利大剧院方案雏形时，寥寥数笔的草图就勾勒出一系列富于感染力的空间景象：有着复杂内部构造而外表纯净的方盒子形体、仪式感强烈的圆柱型入口大厅、多个圆筒交汇形成的富有张力的候场空间、室外水景剧场和屋顶露天剧场等等。时至今日，大剧院几乎是完完整整地以最初构想中的姿态呈现在世人面前，作品完成度之高令人惊叹。
一方面，这体现了建筑师在方案前期对项目全面而周详的考虑，对后期深化实施过程中技术性难题的预见和协调的能力；另一方面，3.65万m²巨大体量的清水混凝土建筑在国内尚属首例，且项目复杂的几何关系和多处大跨度曲面空间导致结构实施难度很大，如何将草图描绘的空间愿景付诸实现，其间的历程可谓艰辛。但安藤先生一直执着于最初的信念，不曾改变初心，对方案精心雕琢至轮廓明朗，再细细打磨出温润的光泽来。这份投入了情感与专注的创造秉承着他一贯的"工匠精神"，也正是这份执着倾注到作品中，才使得建筑传达出直抵人心的感染力。
在与安藤忠雄建筑研究所的合作过程中，我们的团队也一直被这样的

our expertise in the following areas: theatre management, backstage circulation (people and props), broadcasting's stage (and other auxiliary spaces) management, and the simulation of audience experience. Working with the simulations conducted by the acoustics team, we were able to optimise the modelling and elevation of the box and floor seats, allowing for a first-class acoustic experience. In addition, the T-shaped stage design was optimised with the assistance of the lighting, acoustic, and stage mechanics teams. Similarly, all of the professional teams responsible for construction, design, machinery, electricity, lighting, acoustics, interiors and landscape design were encouraged to work as a single force to achieve the vision of the design team. We also played a key role as technical support consultants and project coordinators.

With the unanimous acclaim of our peers in the industry following the first successful performance at the Shanghai Poly Grand Theatre, we also feel greatly honored to have participated in the project's design and construction. We are grateful for all we have learned during our collaboration with a world-class foreign firm. We have not only benefitted from the technical exchanges, but also the collision of thoughts and the pursuit of idealism that reflected in all aspects of work. Each era is defined by its social development, and today's China, just like the choppy river, advances rapidly forward with a huge amount of water and sediment, emitting an impressive roar. Amidst this noise, it is worth maintaining a peace of mind and holding onto one's original ideas and concepts without fear. The only architecture that will remain in people's minds for a long time are those that embrace the difficulty of quality.

(CHEN Jianqiu, QI Xin　Tongji Architecural Design (Group) Co.,Ltd)

精神和情怀感染着。为了能让剧院的主体空间保持最初构想中撼动人心的力量，更重要的是作为一个有着复杂专业技术要求的高品质观演建筑，要保障其各项功能空间得以高效组织、精密运转，让灵动的艺术线条最终得以实现为精准的技术蓝图，我们秉持着专注、严谨的态度贯穿设计和建造过程的始终。

从对剧院管理模式的研究，以及针对演员化妆候场、道具进出、演出转播、剧院管理等辅助空间的安排和流线的组织；到深化剧院内部观众视线的模拟分析，得出最优的观赏效果；配合声学团队的模拟计算，优化池座、楼座的造型和标高从而达到目前国内一流的声学效果；以及配合灯光音响和舞台机械团队的设计，使得"品"字形舞台能够高效地运转；同时整合建筑、结构、机电、舞台机械、灯光音响、室内设计、景观园林等各专业、各设计团队，协同业主为了同一个目标而共同努力，我们在其中起到了技术支撑和协调整合的关键作用。

随着上海保利大剧院的成功首演并得到业内一致的好评，作为参与该项目设计建造团队中的一员，我们也深感荣幸；同时与国际一流的境外事务所的合作过程中我们亦受益良多。这其中不仅仅是技术层面的交流，更为重要的是思想的碰撞，以及通过作品传达出的对理想主义的追求和执着。社会的发展赋予了时代不同的特质，当下的中国就仿佛波涛汹涌的大河，挟带着巨量的水和泥沙飞速地前进，响声震天。在这轰鸣声中，能坚守内心的宁静，始终带着最初的理想与信念不畏艰险一路前行，是值得我们去执着追求的可贵品质。只有在这份品质浸润下设计建造出的作品，才能更长久地存留在人们心中。

[陈剑秋　戚鑫　同济大学建筑设计研究院（集团）有限公司]

---

P214: Architects in construction design stage of studies
P215: Create high quality stage

214 页：建筑师们在施工图设计阶段的探讨研究
215 页：打造高品质的剧院舞台

# Fair-faced Concrete
## 清水混凝土实现

SHI Zhiqiang 史志强 /ZHOU Lanqing 周兰清

In 2010, we made preparations for the project while considering its high degree of structural difficulty as the starting point of all our work. However, despite our preparations, as the engineering work progressed, the challenges we encountered far exceeded our expectations. The most disruptive change for us was the use of unmodified fair-faced concrete, since the moulding systems and processes were totally different from those of modified concrete. Honestly, up until that point, China had no comprehensive or complete process and standards for unmodified fair-faced concrete, meaning that we had no prior experience to draw upon, no previous projects to use as a reference to the work we were doing. We even came close to executing the project under assumption that we were working with domestic (modified) fair-faced concrete. The turning point occurred after the project team visited Japan, during which it witnessed the artistic charm of unmodified fair-faced concrete as presented in Tadao Ando's projects; this experience completely touched and impressed every member of the team, and completely reversed our previous understanding of fair-faced concrete. It gave us the opportunity to really grasp the essence of unmodified concrete. After returning to China, we diligently tackled the technical issues presented by fair-faced concrete via material analysis, conducting dozens of experiments and creating numerous moulding templates. Through this arduous process, we finally worked out the process of creating unmodified fair-faced concrete, determined the optimal concrete mixing ratio, the proper materials for the templates, the ideal construction team, the most suitable concrete supplier, and formulated a series of raw material control measures to prevent variations in the concrete's colour, thereby achieving complete control of the project.

2010年，我们以结构实施难度作为主要切入点开展了各项筹备工作，但即便如此，在工程正式推进过程中所面临的困难和挑战，仍远远超出了我们的预期，其中无修饰清水混凝土定义对于我们来说简直是颠覆性变化，因为修饰和无修饰清水混凝土所要求的模板体系、工艺相差甚远。坦诚地说，中国国内还没有完整意义上的无修饰清水混凝土的工艺和规范，亦无这方面实践经验的积累，在国内我们找不到实体项目样板可供借鉴和参照，所以，我们甚至差点坚持按照国内清水混凝土来组织实施工程。事情的转机出现在项目团队前往日本参观考察之后，安藤忠雄设计的实体工程中无修饰清水混凝土所呈现的艺术魅力，完全打动和折服了考察团队的每一位成员，亦完全颠覆了我们对于清水混凝土的原有认识，让大家有机会真正领会了无修饰混凝土的精髓。

考察回国后，我们坚持不懈地从技术攻关和材料分析工作着手，前前后后经过了数十次的重复试验和样板制作，最终确定了实施工艺方案，择优确定了混凝土配合比、模板材料、劳务队伍、混凝土供应单位等各相关资源，并制定了一系列防止混凝土色差产生的原材料供应管控措施，实现了整个工程质量完全受控。

在整个施工过程中，要保持质量稳定是非常困难的，因为任何一个环节都可能出现差错。我们曾三次要求拆除了存在较大瑕疵的构件，定期检查各个工艺执行情况，我们要求搅拌站为我们项目单独存放砂石料，严格执行我们设计的混凝土配合比，我们的工程师深入一线检查模板加工和安装、混凝土振捣等各个环节。

在历经了三年的坚持和努力后，上海保利大剧院最终以安静、自然而又厚重的美感呈现在世人面前。

The construction site of fair-faced concrete
清水混凝土的施工现场

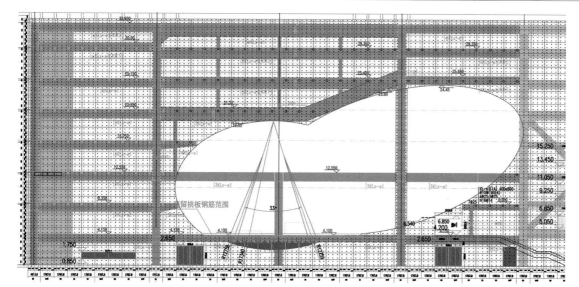

Split design of east facade fair-faced concrete / 东立面清水混凝土分割设计

### Split design
Based on different component types in different sections of the fair-faced concrete structure: the theatre's façade, the arching auditorium walls, concrete beams, and the walls flanking the stairways. The split design was executed, while taking into account of component size, holes position, and decoration.

### 分割设计
针对上海保利大剧院外立面清水墙、观众厅弧形清水墙体、弧形清水梁、直行清水墙、楼梯清水墙体等不同部位、不同类型的构件,结合构件尺寸、洞口位置、后续装饰等进行分割设计。

Template system of ultra-thin steel and wood composite
超薄钢木组合模板体系

### Template design
The new template back pile system, template interlayer switching system, and template pull system were developed by combining the fair-faced concrete split as well as the considerations of its feasibility and economy.

### 模板设计
结合清水混凝土分割,考虑实施可行性与经济性,研发新型的模板背楞体系、模板层间转接系统、模板对拉体系。

It was very difficult to maintain a consistent quality during the whole construction process, for mistakes might arise in any of the steps. We had to order the demolition of the concrete components on three separate occasions due to large defects, and had to check the execution process on a regular basis. We required the mixing station to store the gravel separately for our project, as well as adhere strictly to the concrete mixing ratio that we had determined. Our engineers regularly inspected the construction site to observe and verify the template processing, installation and concrete churning processes, etc.
After three years of persistent effort, the Shanghai Poly Grand Theatre is finally ready for the world, glowing with the quiet, natural, and solemn aesthetic of fair-faced concrete.
The total area of fair-faced concrete in the Shanghai Poly Grand Theatre is about 36,500m². It covers a variety of things: walls, beams, plates, columns, and stairs. In order to perfectly execute the architectural concept, eight months of research and development were carried out, examining topics like split design, template design, fair-faced concrete, and pre-embedment technology. Gradually, we overcame one problem after another and achieved the thick, elegant, and natural style of fair-faced concrete.

上海保利大剧院清水混凝土总面积约3.65万m²,涵盖墙、梁、板、柱、楼梯等多种类型,为完美实现建筑设计理念,我们进行了为期8个月的技术研发工作,从分割设计、模板设计、混凝土研发、预留预埋等技术重点入手,逐渐攻克了一个又一个难题,成功展现了清水混凝土厚重、典雅、朴素自然的风格。

针对大剧院外立面清水墙、观众厅弧形清水墙体、弧形清水梁、直行清水墙、楼梯清水墙体等不同部位、不同类型的构件,结合构件尺寸、洞口位置、后续装饰等进行分割设计,并考虑其实施的可行性与经济性,研发新型的模板背楞体系、模板层间转接系统、模板对拉体系;此外还需对混凝土的配合比和施工工艺进行研发,确保在施工可行的前提下,在混凝土色泽、光泽、裂缝控制等方面做到极致,为此共计进行了600多组小样实验。统计分析清水构件上不同种类的机电、精装修末端、预留洞口,针对不同末端或洞口的工况,结合末端实物样例,制定专项施工措施。

大剧院清水实施工程从2011年开始正式实施,历时13个月基本实施完成,在施工中不断总结前期的经验、优化施工工艺。

在施工过程中,清水混凝土模板体系加工涵盖层间转接托架加工、面板切割、背楞体系加工、对拉螺杆体系加工等多方面,属于模板安装前的准备工作,对保证清水施工质量具有重大意义,而模板安装的好坏直接决定了清水施工的质量,包括模板拼缝错台、水平高差、平整度、顺直度、垂直度等均会直接影响混凝土实施效果。

Sample wall test and mixing ratio development sample
样板墙试验及配合比研发小样

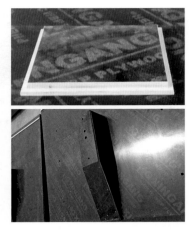

Embedded plate and case of fair-faced concrete
清水混凝土预埋衬板与预埋盒体

On-site construction jobs
现场施工作业

Flatness check, abnormal keel processing
平整度校核、异形龙骨加工

Based on different component types at different sections of the Grand Theatre's façade, arching auditorium concrete walls, arching concrete beams, vertical walls, and stairway walls, we decided to execute the split design, with consideration of the following elements: component size, hole position, decoration, template back pile system, template layer switching system, and template pulling system. All of the elements were considered with regards to their feasibility and economy. In addition, we researched and developed concrete mixing ratio and creation technologies to ensure the perfect concrete colour and lustre, as well as to ensure its structural integrity and feasibility. Over 600 groups of sample experiments were conducted to achieve this result. Statistical analyses were conducted for different types of motorized electricity, terminals, and drilling holes on fair-faced concrete components. Special construction measures were prepared according to the different conditions of different terminals or holes and the physical terminal samples.

The fair-faced concrete project of the Grand Theatre was formally executed in 2011, and was completed after thirteen months. During this time, we constantly consulted our previous experience to develop optimal construction techniques and technologies.

During the construction process, the fair-faced concrete template processing system consisted of the following: interlayer bracket machining, surface plate cutting, back pile system processing, and pull screw system processing, all of which were preparatory work preceding template installation. The system factored significantly into the creation of quality fair-faced concrete. The quality of the template installation directly determined the quality of the resultant fair-faced concrete, i.e., the template's seaming, (staggered) slab arrangement, level differences, flatness, and angles would all affect the resultant fair-faced concrete. A careful concrete pouring process was also required for good template installations; otherwise, the formation of honeycombs, pitting surfaces, and excessive/insufficient amount of vibratos would render all our previous efforts for naught. Given the irreversibility of changes applied to fair-faced concrete components, the protection of finished products was of utmost importance. Effective protective measures would reduce the pollution and destruction caused by the ensuing construction or cross operations; therefore, successful creation of fair-faced concrete was inseparable from efficient management, requiring additional detailed planning, meticulous execution, and timely extraction.

The first consisted of standardized management and operation processes, whose operations, inspection methods, feedback and closing mechanisms were based on previous successful experiences. The second was details-oriented management, drawing heavily upon existing data, strict quality control, and "100% check" of all materials of processes; this ensured high standards in management and project execution. Finally, there was simulation-based management, whose purpose was to reduce the number of potential blind spots in management. Here, BIM technology was applied before the execution of every fair-faced concrete project, simulating the results of the process, identifying potential problems that were previously overlooked, and making the proper adjustments.

(SHI Zhiqiang, ZHOU Lanqing  China State Construction Engineering Co., Ltd Shanghai)

The cylindrical space under construction / 圆筒空间施工现场

Operation training wall / 操作训练墙

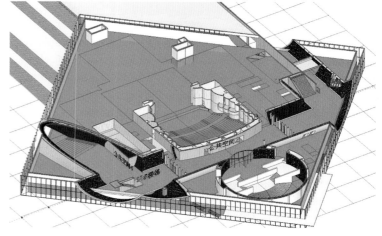

BIM technology simulation parts and components / BIM 技术模拟部位图和构件图

100% torque check / 100% 力矩检查

良好的模板安装还需要精心的混凝土浇筑施工，否则，浇筑引起的蜂窝麻面、过振少振等质量缺陷足以让前期的辛劳付诸流水。鉴于清水混凝土构件的不可逆性，成品的保护就显得格外的重要。采取有效的保护措施能有效地减少后续施工或者交叉施工引起的污染以及破坏；因此，清水混凝土的成功实施与高效管理是分不开的，它需要更细致的策划、更深入的执行和更及时的提炼。首先是标准化管理；基于前期成功的经验积累，形成标准化的管理流程、操作方式、检查方式、反馈机制以及总结机制。其次是精细化管理；参数化的材料管理、精益求精的质量管理、100% 的检查模式无不体现着管理的高执行力、高贯彻落实要求。最后是仿真化管理；为了提高管理力度，降低管理盲区，在清水工程分段正式实施前，采用BIM 技术进行施工模拟，预演整个施工过程，及时发现考虑不周之处，并进行相应优化。

[ 史志强 周兰清　中国建筑股份有限公司（上海）]

# Interior Decoration
室内装饰实现

LAN Hai 兰海 / WANG Jian 王键

Shanghai Poly Grand Theatre is a project connecting the spaces of the theatre complex via the intersection of "cylindrical" volumes, which happens to also link the structure with its natural surroundings. The interior decoration of the project is very pure and concise, with black granite for the floors, wooden grille for the cylindrical forms, and a small amount of fair-faced concrete for the walls. In order to highlight these features, extreme attention to detail is needed to ensure their overall quality. This is the most important factor in the project's overall quality, as well as the main reason for its high difficulty of execution. First of all, the "cylindrical" components were the biggest difficulty in interior decoration, but also its key highlight, a real feat of engineering in the construction of public spaces and a masterful design concept. The installation of the large wooden grille surfaces was the biggest problem that plagued us and the clients at the time. Mr. Ando had wanted to use natural wood, but that failed to materialize due to differences in fire safety standards between China and Japan. After lengthy research and discussion, we decided to use aluminum grilles, using thermal transfer printing technology to apply natural wood grain effect on their surface. As for the treatment of the cylindrical volumes, there were numerous correlation and connection problems between the internal cylinder and other spatial openings, such as the connection with the escalator entrance, the connection with the indoor opening, the connection with the corridor, and the connection with the ground-floor walls. The treatment methods applied to these linkages hinged were dependent on the overall quality of the interior construction. A meticulous handling of these spaces required a lot thought, precision, and accuracy during the entire construction process.
Secondly, the decoration of the auditorium was the core of the entire project, an important space integrating artistry and functionality, meaning that the installation had to account for the acoustics of the architecture, the stage equipment, and the needs of the audience. While priority was given to functionality, artistic and decorative requirements were also met to the greatest extend possible. Most notable was the decoration of the auditorium walls, which was conducted with utmost difficulty, as the undulating wooden grilles were applied to the walls to

上海保利大剧院是一个空间项目，通过"圆筒"使建筑空间得以相互贯通，同时也通过它使建筑与自然环境相通相映；而这个项目的室内装饰形式也是极为简洁、纯粹的——地面为黑色的花岗岩，圆筒装饰全部为暖色的木质格栅，同时贯穿以少量的清水混凝土墙面。为了很好地表现这些特点，就需要通过极其精造的细节来体现整体的品质，细节是体现这个项目品质最为重要的因素，同时也是这个项目实施难度的最主要原因。

首先，"圆筒"的装饰施工是内装项目中最大的难点，同时也是重点和亮点。因为这是公共区域装饰中最大的一项，同时也是设计概念的集中体现；其中大面积木格栅装饰面的工艺实现，是一直困扰我们及业主方的最大问题。安藤先生想要用天然木质来实施，但由于国内消防标准的不同，而未能实现。后来经过反复的研究与探讨，最终我们只能采用铝合金型材作为格栅本体，表面采用热转印技术模仿天然木纹效果的工艺方法来实现。另外，这个项目中存在很多圆筒内部与其他洞口的关联和交接问题，如与扶梯洞口的交接、与室内洞口的交接、与连廊的交接、与墙地面的交接等，这些细节如何处理，成败关乎到整体内装施工的品质。如何把这些部位处理好，是要面临很大的难度，同时也要花费很多的心思，对施工的精细度和精准度都有很高的要求。

其次，观众厅的装饰施工是整个项目的核心，这是一个集艺术性与功能性于一体的重要空间，因此，它的实施还与建筑声学、舞台设备及观众使用等功能需求密切相关。在满足功能性的前提下，再尽最大可能地满足装饰性和艺术性的需求。其中难度最大的是墙面施工，按照设计图效果要求，观众厅的墙面需采用凹凸错落的木制装饰，把这里打造成为一个传递声音的"乐器"。因此在业主方的主导和建议下，我们采用经过特殊防火处理50mm厚的松木集成材，将其切割成不同形状的弧型板并打磨光滑面饰清漆，凹凸错落地安装在墙上。这一做法，既满足了装饰性的要求，同时也最大限度地满足了声学和消防等方面的要求。完工后的观众厅墙面散发出如同乐器般柔和高雅的光泽，其整体声学测试空场混响时间，是目前全上海所有剧院中，建声条件最好的。

最后，大剧院的照明设计是安藤大师巧妙运用自然光塑造建筑形象的成功案例之一。在与安藤建筑事务所多次沟通后，最终确定以简洁、单纯为原则来展现、烘托建筑空间，其设计主要包括两点。第一，室外泛光设计：在建筑玻璃幕墙和清水混凝土外墙之间，以层间格栅为单元设洗墙灯具，将清水混凝土墙面整体均匀照亮。经计算机软件模拟及灯光试验确定灯具选型为5,000K色温的LED条形灯具；建筑泛光的场景设计上采用减法原则，放弃了绚丽多彩的光色及过多灯光变化模式，以更好地契合其观演建筑的

Preliminary grille template
格栅初步样板

Mid-term grille template
格栅中期样板

Post grille template
格栅后期样板

Core insertion method for grille junctions
格栅连接处的插芯做法

meet design requirements stipulating the walls to become a "musical instrument" for the human voice. Therefore, under the chief architect's guidance and suggestion, a 50mm integrated pine timber, processed by a special fire prevention treatment, was chosen, cut into different arching plates, polished and varnished, and finally installed onto the walls. This method not only satisfied the decorative requirements, but also, to the greatest extent possible, the acoustic and fire safety requirements. Upon completion, the auditorium wall exuded a smooth and elegant lustre, just like the surfaces of musical instruments. And its acoustic frequency results during the overall acoustics test showed the venue to possess excellent acoustics, making it one of the best theatre venues in modern Shanghai.

Finally, the lighting design of the Grand Theatre was one of Mr. Ando's successful ideas for the structure: to create an image of the architecture by clever application of natural light. After lengthy communication with Tadao Ando Architect & Associates, we agreed to apply the principles of purity and concision to the interior space. The design consisted of two main points. First, there was the outdoor floodlighting design, which was located between the building's glass curtain wall and its external fair-faced concrete wall. The wall lights were arranged by using the interlayer grilles as the organising unit, providing uniform illumination of the entire concrete wall. Through computer simulations and lighting tests, LED lights with a color temperature of 5,000K were chosen; the subtraction principle was later applied to match the tone of the lighting with the purpose of the venue, giving up some of the brighter colors and reducing the number of lighting patterns to harmonize with the venue's purpose as a dramatic performance space. Second, there was the spatial lighting design for the indoor and outdoor cylindrical forms. In order to execute the overall lighting principle of "lighting without a lamp," the main lights in the cylindrical area were hidden in the steel guardrails at the bottom of the cylinder. The final lighting effects on the cylinders, arcs that change according to the illumination from below, then, would be determined by the lighting angle, level of luminosity, and degree of uniformity. The 2,700K LED wall lights were chosen after conducting over twenty on-site lighting tests (with the LED lens angle set at 85°, with a polarization of 20°). After installation, nearly the entire cylinder was illuminated, with lighting being brighter near the bottom and gradually becoming dimmer as it approaches the top. The warmer texture of the 2,700K color temperature was also at accentuating the warm feeling exuded by the walls' wooden surface. The sharp contrast formed between the warm wooden walls and the cold fair-faced concrete gave the building a brand new visual image. It is worth mentioning that at the outdoor space located at the eastern side of the complex, the intersection between the horizontal cylinder and the cylinder tilted at a 15° angle posed huge difficulties during construction. While the intersection of the two cylinders created an irregular spatial arrangement, it was nevertheless necessary to find a way to illuminate the areas in a uniform manner, employing the principle of "lighting without a lamp." The ideal lighting effects were finally achieved by making use of combined use of lighting fixtures (which were installed earlier), and other types of lighting cast at horizontal, 30° and 60° angles, etc.

(LAN Hai, WANG Jian  Beijing Qingshang Architectural Ornamental Engineering Co., Ltd.)

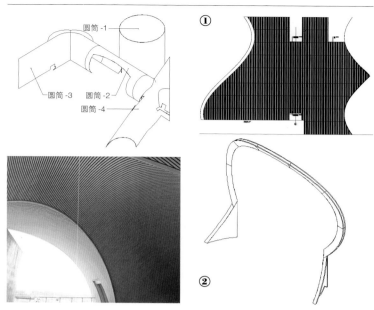

**Treatment of special parts**

The grille at the cylinder's intersection nodes varied in length and in angle; therefore, it was difficult to measure the irregular grille sizes onsite, and on a one-by-one basis. 3D software was relied upon to carry out the 3D modelling for each cylinder, taking into the relative positions and tilt angles which were already known. Each model surface was spread out, the grille design drawn on, and the angles and shapes of the various grille parts calculated.

Figure 1 is the developed grille diagram for Cylinder-3. The grille specifications at the intersections could be obtained from the chart. Configurations for other grilles can be obtained in the same manner. This process allowed for the conversion of spatial dimensions into dimensions on a flat plane, greatly improving cutting precision and reducing the amount of grille wasted.

The aluminium plating was needed link the cylinders at their intersections. The plating was not the typical plate cast in a hyperbola, but was designed via the distortion of a single plate, as shown in Figure 2 (for example, the aluminum plate at the intersection of Cylinder-1 and Cylinder-2). Moreover, different sections of each aluminum plate had different widths, increasing the difficulty of product design and processing. Although these aluminum plates were not often used in the project, they were all situated in difficult-to-access locations.

**特殊部位的处理**

圆筒在相交相贯部位格栅的长短是不一样的，每一端的角度也是不确定的，现场很难逐一将这些不规则的格栅尺寸测量出来，所以就依靠三维软件将各个圆筒根据相对位置、倾斜角度等已知条件进行三维建模，利用三维软件将模型表面进行展开，然后在展开平面图形上等比绘制格栅分格图、格栅收口部位的角度及形状示意图。图1为圆筒-3展开分格图，依此展开图及格栅排布，可得出收口部位的格栅规格，其他圆筒同理操作。通过这些转化，使得空间尺寸变为平面尺寸，大大提高了下料的准确性和减少了格栅的损耗率。

圆筒与圆筒相交接部位需要铝板进行收口，这种收口铝板不是普通概念的双曲板，而是单曲的扭曲形式，如图2（以圆筒-1与圆筒-2相贯部位收口铝板为例）所示可以看出铝板的扭曲形式，而且不同部位的铝板之间的角度与板面宽度均是不同的，对于产品设计与加工要求比较高，虽然这种铝板在整个项目中数量不多，但却都处在难度较高的部位。

文化气质。第二，室内及半室外圆筒空间照明设计：为了贯彻整体照明设计"见光不见灯"原则，圆筒区域主照明暗藏于圆筒底部的钢质护栏内。由于灯光从下部照射会形成圆筒表面的弧度变化，因此对灯光照射角度、照度、均匀度的控制决定最终的效果。通过二十余次的现场光效试验、对比，最终确定采用订制2,700K色温的条形LED光源偏光洗墙灯具（LED透镜角度为85°、偏光20°）。安装完成后，灯光照射区域基本覆盖了整个圆筒，从下至上由强到弱的渐变退晕光效，很好地突出了圆筒的形体特征。2,700K色温的暖色调光源也更好地烘托出圆筒壁木纹格栅温暖的质感，与清水混凝土的冷白色温形成强烈的对比，赋予建筑崭新的视觉形象。

值得一提的是建筑东侧室外开放空间15°倾斜圆筒与水平圆筒的相交处，因倾斜、切割出的空间为异形不规则空间，同时又受"见光不见灯"的原则所限，要保证照度均匀、与其他圆筒开口处相平衡的照明效果有很大的难度。为此，利用建筑有限的预留灯位，结合使用了条形洗墙、30°投光、60°投光等多种类型灯具，最终取得了理想的照明效果。

［兰海 王键 北京清尚建筑装饰工程有限公司］

# Stage Mechanics
舞台机械实现

LIU Jianbin 刘建斌

The stage of the Grand Theatre applied the standard T-shaped stage design. The setup includes a main stage, left and right side stages, and a backstage. The proscenium opening is 16m wide and 11m high. The main stage is 30m wide and 22m deep, while the stage grille is 25.6m high. Five sets of dual lifting platforms are arranged in the main staging area, each with a height of 4.5m. On the left and right side stages, there are five sets of stage trolleys, five sets of auxiliary lifting platforms, and five sets of compensation lifting platforms. In the backstage, a turntable allows for forward and backward motion, and is complemented by a set of auxiliary lifting platforms and five sets of compensation platforms. Two sets of orchestra lifting platforms are set in the proscenium opening, which can extend the main staging area forward, thereby increasing performance space.

The lifting platform is the most important part of the modern mechanical stage, as well as its most important piece equipment. It allows for the rich manipulation and transformation of the stage, enabling complete usage of the space encompassed by the stage plane. Different combinations of lifting platforms allow for the creation of different performance spaces. For instance, the central lifting platform can be lowered to a certain height, while the side stages can be raised, creating changes in scenery and allowing the stage itself to participate in the performance. In the backstage, the turntable set can be moved to different positions on the stage to participate in the spectacle, and the main turntable can rotate both clockwise and counterclockwise to generate unique visual effects.

Eight single-lift cranes and one subtitle screen are arranged above the auditorium's proscenium opening. Above the main stage area, one can observe the following: proscenium lighting device, fire curtain, curtain machine, a false proscenium opening, primary and secondary screens, five lighting suspenders, fifty three electric suspenders, and six single-lift cranes. At both sides of the stage, there are two electric booms, and two sets of light hangers. Two sets of rail cranes are installed for the left and right stages, and are used to move heavy stage decorations and equipment. In the backstage there are thirteen dual-purpose landscape light booms. Finally, the stage also features a movable sound hood. The equipment above the stage allows for flexible stage setups, diversifying performance content and effects. The installation of the proscenium lighting device is an innovation on stage lighting, as this design has never been done before in China.

The multipurpose hall in the Poly Grand Theatre is intended for small conferences and performances. Equipment below the stage includes: a main lifting platform, scissor lift, and lifting platforms for the forward areas. Stage equipment includes dual-purpose landscape light suspenders, side light hangers, single-lift cranes and rail cranes. By applying different combinations of the lifting platform, adjustable and temporary seating, the multipurpose hall space can be arranged into

大剧场的舞台结构形式采用典型的主舞台、左右侧台、后舞台的"品"字型工艺布置，台口宽16m，高11m；主舞台宽度30m，进深22m，栅顶高度25.6m。

主舞台区域共设置有5套升降台，升降台为双层结构，层高4.5m；在左右侧台各设置有5套侧台车台、5套辅助升降台和5套补偿升降台；后舞台设置可前后移动的车载转台、1套后辅助升降台以及5套后台补偿台；台口前设置2套乐池升降台，可以使主舞台区域向前延伸，增加演出空间。

升降台是现代化机械舞台的主体，是台下舞台设备最重要的组成部分，它能够灵活、丰富地变换舞台形状，使整个主舞台在平面和台阶之间变换。通过升降台的相互组合，改变升降高度，可形成不同的演出空间；升降台下降一定高度后，侧台车台可以运行到升降台上，并可随之升降，既可以迁换布景，又可以参与演出。设置于后舞台的车载转台可以移动到主舞台的不同位置参与演出，中心转台可以在行走过程中正向或反向转动，产生独特的视觉效果。

大剧场的台口外上空设置了8台单点吊机，1道字幕屏吊杆；主舞台区域上空依次布置台口灯光装置、防火幕、大幕机、活动假台口、二幕机、5道灯光吊杆、53道电动吊杆和6个自由单点吊机；舞台两侧设置了2道侧电动吊杆和2套侧灯光吊架；左右侧台上空分别配置了4套轨道吊机，用于移动重型道具和装台；后舞台上空设13道景灯两用吊杆。此外舞台内还设置了1套移动拼装式反声罩。

舞台上空设备可根据需要任意组合使用，以提高舞台布景布局的灵活性，丰富表演内容和效果。其中，台口灯光装置的使用突破了国内剧院灯光布局的既有模式，效果良好。

---

P222: The stage machinery and equipment
222 页：舞台机械设备

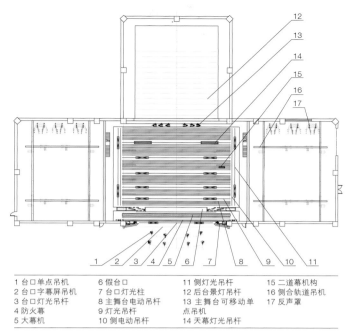

| | | | |
|---|---|---|---|
| 1 台口单点吊机 | 6 假台口 | 11 侧灯光吊杆 | 15 二道幕机构 |
| 2 台口字幕屏吊杆 | 7 台口灯光柱 | 12 后台景灯吊杆 | 16 侧合轨道吊机 |
| 3 台口灯光吊杆 | 8 主舞台电动吊杆 | 13 主舞台可移动单 | 17 反声罩 |
| 4 防火幕 | 9 灯光吊杆 | 点吊机 | |
| 5 大幕机 | 10 侧电动吊杆 | 14 天幕灯光吊杆 | |

Layout plan of Grand Theatre stage equipment
大剧场台上设备平面布置图

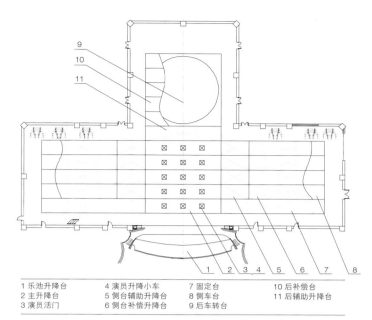

| | | | |
|---|---|---|---|
| 1 乐池升降台 | 4 演员升降小车 | 7 固定台 | 10 后补偿台 |
| 2 主升降台 | 5 侧台辅助升降台 | 8 侧车台 | 11 后辅助升降台 |
| 3 演员活门 | 6 侧台补偿升降台 | 9 后车转台 | |

Layout plan of Grand Theatre equipment under stage
大剧场台下设备平面布置图

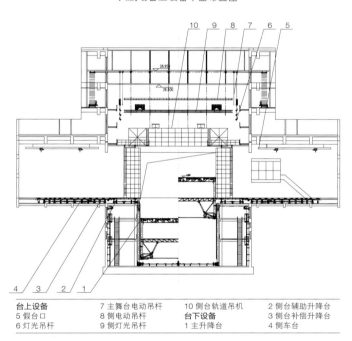

| 台上设备 | 7 主舞台电动吊杆 | 10 侧台轨道吊机 | 2 侧台辅助升降台 |
|---|---|---|---|
| 5 假台口 | 8 侧电动吊杆 | 台下设备 | 3 侧台补偿升降台 |
| 6 灯光吊杆 | 9 侧光吊杆 | 1 主升降台 | 4 侧车台 |

Transverse plan of Grand Theatre equipment arrangement
大剧场设备布置横剖图

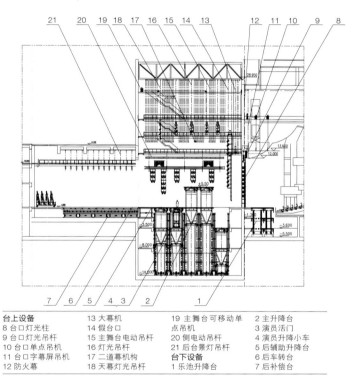

| 台上设备 | 13 大幕机 | 19 主舞台可移动单 | 2 主升降台 |
|---|---|---|---|
| 8 台口灯光柱 | 14 假台口 | 点吊机 | 3 演员活门 |
| 9 台口灯光吊杆 | 15 主舞台电动吊杆 | 20 侧电动吊杆 | 4 演员升降小车 |
| 10 台口单点吊机 | 16 灯光吊杆 | 21 后台景灯吊杆 | 5 后辅助升降台 |
| 11 台口字幕屏吊杆 | 17 二道幕机构 | 台下设备 | 6 后车转台 |
| 12 防火幕 | 18 天幕灯光吊杆 | 1 乐池升降台 | 7 后补偿台 |

Longitudinal plan of Grand Theatre equipment arrangement
大剧场设备布置纵剖图

a diverse array of settings: open stage, closed stage, conference hall, T-shaped stage, and centre stage (i.e. stage surrounded by rings of seats), satisfying various performance and spacing needs.

System control of the stage machinery consists of a distributive control structure and a specialised control software to fulfill operative and monitoring functions. The setup allows for an effective operation interface, introducing predictive algorithms to detect and adjust stage equipment as needed, thereby ensuring a safe environment for both personnel and the machinery. An excellent human-system interface allows for clear and simple operation, as evidenced by its use of the digital graph display, an easy-to-operate menu screen, automatic memory, fault-processing system, and interlocking information systems, etc.

(LIU Jianbin  General Armament Department Engineering Design & Research Institute)

剧院的多功能厅用于小型会议和演出。台下设备包括主升降台、子母升降台、前区升降台；台上设备包括灯景两用吊杆、侧灯光吊架、单点吊机和轨道吊机等。通过升降台、伸缩座椅和临时座椅的组合使用，能够实现伸出式舞台、尽端式舞台、会议模式舞台、"T"形台和中心式舞台等不同的舞台模式，以满足不同演出需要。

剧院的控制系统采用分布式控制结构和专用控制软件，对所有舞台机械的驱动装置实施运行控制和状态监视，并提供操作界面和操作方法；对安全相关设备引入动态运行目标检测、互锁预判算法，保证设备及人员安全。系统具有良好的人机界面，操作简单明确，并具有图形数字显示、屏幕菜单操作、自动记忆、故障处理、互锁信息提示等功能。

[刘建斌  总装备部工程设计研究总院]

# Data
## 数据

**Architecture Name:** Shanghai Poly Grand Theatre
**Address:** No. 159, Baiyin Road, Jiading District, Shanghai, China
**Client:** Shanghai Poly Maojia Real Estate Development Co., Ltd.
**Function:** Theatre.
Mostly for operas, symphonies and large-scale theatrical performances. Also for conferences, celebrations and ceremonies.

### Design · Supervision
**Architect:** Tadao Ando Architect & Associates
Tongji Architectural Design (Group) Co.,Ltd
**Schematic Design:** Tadao Ando Architect & Associates/Tadao Ando, Kazuya Okano, Yoshinori Hayashi
**Executive Architect/Engineer:**
**General:** Tadao Ando Architect & Associates/Tadao Ando, Kazuya Okano, Yoshinori Hayashi
Tongji Architectural Design (Group) Co.,Ltd/CHEN Jianqiu, QI Xin, ZHANG Rui, TANG Yanli
**Interior:** Tadao Ando Architect & Associates/Tadao Ando, Kazuya Okano, Yoshinori Hayashi
Beijing Qingshang Architectural Ornamental Engineering Co.,Ltd/SU Liqun, LAN Hai, JIANG Fengze, WANG Jian, WANG Zhiyong
**Structure:** Tongji Architectural Design (Group) Co.,Ltd/LIN Jianping
**MEP:** Tongji Architectural Design (Group) Co.,Ltd/TAN Limin, WEI Jiancheng, YAN Zhifeng, LIU Jin
**Landscape:** Tadao Ando Architect & Associates/Tadao Ando, Kazuya Okano, Yoshinori Hayashi
Tongji Architectural Design (Group) Co.,Ltd/LU Weihong, FENG Bonan
**Acoustics:** ZHANG Kuisheng Acoustical Design and Research Studio/ZHANG Kuisheng, SONG Yongmin (Karasawa Architectual & Acoustic Design)
**General Coordinator:** CA-GROUP/MA Weidong, LI Yi, CHEN Xudan
**Construction Manager:** Shanghai Poly Maojia Real Estate Development Co., Ltd./SUN Jian, GONG Zhigang, LIANG Feng, WEI Shixiong, ZHAN Yunliang, PAN Lican
**Supervision:**
Shanghai Poly Maojia Real Estate Development Co., Ltd. (General supervision)
Shanghai Construction Supervision Consulting Co.,Ltd (The third party)/SUN Yalong, LI Xiubing
Tadao Ando Architect & Associates (Design supervision)

**Consultants:**
Acoustical Consultant: Shanghai Yongjia Electronic Engineering Co.,Ltd
Stage Consultant: General Armament Department Engineering Design & Research Institute (ZHENG Zhirong, LIU Hailin, LIU Jianbin, MA Tianzhou, ZHANG Liping, CHEN Wei) Curtain Consultant: Shenzhen China Aviation Curtain Wall Engineering Co.,Ltd. (LI Taiyi)

### Construction
**General/ Civil Work:** China State Construction Engineering Co., Ltd Shanghai/ZHOU Lanqing, HAN Guoqiang, ZHANG Zhongliang, CHEN Li, XIE Yude
**Electrical:** China Railway Construction Group Co. Ltd
**HVAC:** China Railway Construction Group Co. Ltd
**Water Supply and Drainage Engineering:** China Railway Construction Group Co. Ltd
**Interior:** Beijing Qingshang Architectural Ornamental Engineering Co.,Ltd
**Installation:** Huading Architectural Décor Engineering Co.,Ltd
**Exquisite Decoration:** Beijing Qingshang Architectural Ornamental Engineering Co.,Ltd
Shenzhen Zhongfutai Cultural Architecture Co.,Ltd
Dalian Qiansen Sport Facility Engineering Co.,Ltd
Daiichi International (Holdings) HK., Ltd
**Landscape:** Shanghai Chunqin Garden Engineering Co.,Ltd/SONG Xingqin, FENG Ji
**Sound System:** Shanghai Yongjia Electronic Engineering Co.,Ltd
**Stage:** General Armament Department Engineering Design & Research Institute
**Mechanical and Electrical Installation** China Railway Construction Group Co., Ltd
**Curtain Wall:** Shenzhen Curtain Wall Co., Ltd
**Fire:** Shanghai Zhiyuan Security Fire Protection Engineering Co.,Ltd/GU Wei, ZHANG Zhen )
**Lighting:** Taiji Computer Corporation Limted
**Electrical power:** Shanghai Weiming Power Equipment Co.,Ltd

### Scale
**Site Area:** 30,235m²
**Building Area:** 12,500m²
**Total Floor Area:** 55,904m²
**Ground Floor Area:** 35,840m²
**Capacity Area:** 30,730m²
**Underground Floor Area:** 20,064m²
**Floor Area Ratio:** 1.02
**Building Coverage:** 41.34%
**Green Ratio:** 25%
**Concentrate Green Space Ratio:** 15%
**Number of Floors:** Underground: 1, Ground: 6
**Area of Each Floor:** BF: 18,173m²/1F: 12,112m²/2F: 3,062m²/3F: 3,157m²/4F: 3,850m²/5F: 4,896m²/6F: 2,108m²
**Height:**
**Building Height:** 34.4m
**Floor Height:** No Typical Floor 4.2m/6.3m
**Ceiling Height:** 3~30m
**The highest Eaves Height:** 33.2m
**Parking:** 336 cars
**Ground:** 19 (Bus: 7, Car: 12)
**Underground:** 317
**Non-motor Vehicles:** 143

### Project Period
**Design Period:** 2009.5-2010.10
**Construction Period:** 2011.3-2014.8

### Site Information
**Road Width:** West Side 16m, North Side 35m

### Structure
**Main Structure:** Concrete Structure, Steel Reinforced Concrete Structure
**Part of the Structure:** Steel Frame Structure
**Foundation:** PHC-pile
**Anti-seismic:** 7 Degree Seismic Fortifications

### Equipment
**Air Conditioning System:**
Heat source: atmospheric gas-burning boiler
Air conditioning system: cold source /ice storage air conditioning system
Tail end/air processer and fan coil
**Electrical System:**
Power supply mode: double circuit
Equipment capacity: total transformer capacity: 6,250kVA
Among them, 2 sets of 2,000kVA equipment, 1 set of 1,250kVA and 1 set of 1,000kVA
Rated voltage: 10kV±5%/0.4kV
Backup power supply: standby diesel generator set 1,200kW
**Sanitary Equipment:**
Water supply: municipal water supply, recycled rainwater supply
Hot water supply: local water supply form
Water drainage: siphon drainage
Catch basin: rainwater recovery pond
**Disaster Prevention Equipment:**
Fire: Indoor fire hydrant, Fire sprinkler
Smoke Exhaust System: Mechanical and natural ventilation
Elevator: 13 Lifts + 1 Escalator

### External Finish
**Roof:** Asphalt waterproofing with insulation, Planting, Stone-Paving, Wood deck on cinder concrete layer
**Wall:** Fair-faced Concrete Wall+Aluminum Curtain-Wall(clear laminated glass)
Opening: Aluminum sash(Low-e paired glass)
Aluminum profile rib-wall(wood style paint finish)
**Landscape:** Greenery/Beech, Camphor, Prunus subhirtella, Cryptomeria japonica, Sweet scented osmanthus, Ginkgo biloba Ground/Granite, Pool/Pebbles

### Interior Finish
**Entrance Hall:**
**Ceiling:** Mineral Wool board, Plaster board with paint
**Wall:** Aluminum profile rib-wall(wood style paint finish)
**Ground:** Granite paving
**Foyer:**
**Ceiling:** Aluminum profile rib-wall(wood style paint finish)
**Wall:** Aluminum profile rib-wall(wood style paint finish)
**Fair-faced Concrete:**
**Ground:** Granite paving
**Auditorium:**
**Ceiling:** GRG panel with paint
**Wall:** Laminated-wood rib-wall(heated with fire proof agent)
**Ground:** high-strength solid wood composite floor
**Other Functional Space:** Light steel keel gypsum board decoration for walls and ceilings

### Energy Saving
STP Thermal Insulation Board (roof and walls)

建筑名称：上海保利大剧院
地址：上海市嘉定区白银路 159 号
业主：上海保利茂佳房地产开发有限公司
用途：剧院
主要承接各种歌舞剧、戏剧、交响乐和大型综合文艺演出，并具备举行大型会议或庆典活动的功能

设计・监理
设计方：安藤忠雄建筑研究所
　　　　同济大学建筑设计研究院（集团）有限公司
方案设计：安藤忠雄建筑研究所
　　　　　安藤忠雄 冈野一也 林庆宪
施工图设计：
　建筑：安藤忠雄建筑研究所
　　　　冈野一也 林庆宪
　　　　同济大学建筑设计研究院（集团）
　　　　有限公司 / 陈剑秋 戚鑫 张瑞
　　　　汤艳丽
　内装：安藤忠雄建筑研究所 / 安藤忠雄
　　　　冈野一也 林庆宪
　　　　北京清尚建筑装饰工程有限公司 /
　　　　宿利群 兰海 江凤泽 王健 王志勇
　结构：同济大学建筑设计研究院（集团）
　　　　有限公司 / 林业平
　设备：同济大学建筑设计研究院（集团）
　　　　有限公司 / 谭立民 韦建成
　　　　严志峰 刘瑾
　景观：安藤忠雄建筑研究所 /
　　　　冈野一也 林庆宪
　　　　同济大学建筑设计研究院（集团）
　　　　有限公司 / 陆伟宏 冯博楠
　声学：章奎生声学设计研究所 / 章奎生
　　　　宋拥民（扩初设计时由唐泽诚建筑
　　　　音响设计事务所配合）
综合项目管理：文筑国际 / 马卫东 励懿 陈旭丹
业主方项目管理：上海保利茂佳房地产开发有
　　　　限公司 / 孙健 龚志刚 梁锋
　　　　魏世雄 詹云亮 潘立灿
监理：上海保利茂佳房地产开发有限公司（综合监理）
　　　　上海建科监理咨询有限公司（第三方监理单位）/ 孙亚龙 李骜兵
　　　　安藤忠雄建筑研究所（设计方监理配合）
分项顾问单位：
　音响顾问：上海永加电子工程有限公司
　舞台顾问：总装备部工程设计研究总院
　　　　郑志宋 王海林 刘建斌
　　　　马天舟 张丽萍 陈威
　幕墙顾问：深圳中航幕墙工程有限公司
　　　　李太义

施工单位
总包：中国建筑股份有限公司（上海）
　　　周兰清 韩国强 张忠良 陈立 谢玉德
电气：中铁建设集团有限公司
暖通：中铁建设集团有限公司
给排水：中铁建设集团有限公司
内装：北京清尚建筑装饰工程有限公司
初装：华鼎建筑装饰工程有限公司
精装：北京清尚建筑装饰工程有限公司
　　　深圳中孚泰文化建筑建设股份有限公司
　　　大连千森体育设施有限公司
　　　香港成功国际集团有限公司
景观：上海春沁园林工程建设有限公司 /
　　　宋兴琴 冯吉
音响：上海永加电子有限公司
舞台：总装备部工程设计研究总院
机电安装：中铁建设集团有限公司
幕墙：深圳中航幕墙工程有限公司
消防：上海智源安保消防工程有限公司 /
　　　顾玮 张震
弱电：太极计算机股份有限公司
电力：上海为明电力设备有限公司

规模
用地面积：30,235 ㎡
建筑底层占地面积：12,500 ㎡
建筑面积：55,904 ㎡
地上：35,840 ㎡，计容面积：30,730 ㎡
地下：20,064 ㎡
建筑密度：41.34%
容积率：1.02
绿地率：25%
集中绿地率：15%
各层面积：
地下 1 层：18,173 ㎡ / 1 层：12,112 ㎡ / 2 层：3,062 ㎡ / 3 层：3,157 ㎡ / 4 层：3,850 ㎡ / 5 层：4,896 ㎡ / 6 层：2,108 ㎡
高度：
建筑高度：34.4m
层高：无标准层 4.2m、6.3m
天花高：3~30m
最高挑高：33.2m
机动车停车位：336 辆
地上：19 辆（大客车 7 辆，小轿车 12 辆）
地下：317 辆
非机动车停车位：143 辆

项目时间
设计时间：2009 年 5 月~2010 年 10 月
施工时间：2011 年 3 月~2014 年 8 月

基地条件
道路宽度：西面 16m，北面 35m

结构
主体结构：混凝土结构、钢筋混凝土结构
一部分：钢结构
桩基：PHC 桩
抗震：7 度设防

设备
空调设备
热源：常压燃气锅炉
空调形式：冷源 / 冰蓄冷空调系统
末端 / 空气处理机加风机盘管
电气设备：
供电方式：两路供电
设备容量：变压器总容量：6,250kVA
其中 2 台 2,000kVA，
1 台 1,250kVA、1 台 1,000kVA
额定电压：10kV ± 5%/0.4kV
后备电源：备用柴油发电机组 1,200kW
卫生设备：
给水：市政给水、雨水回收再利用 加压给水形式
供热水：局部供水形式
排水：虹吸排水
集水槽：雨水回收水池
防灾设备：
消防：室内消防栓、喷淋
排烟：机械排烟 + 自然排烟
电梯：
13 台竖向电梯 +1 台自动扶梯

外部完成面
屋顶：沥青隔热防水 + 煤渣混凝土之上种植
屋面、青石板走道、木平台
外墙：清水混凝土 + 铝框幕墙（超白玻璃幕墙）
开口部：铝板（Low-e 复合玻璃）+ 木纹铝格栅
景观：绿化、榉树、樟树、日本早樱、日本柳杉、桂花、银杏等
地面 / 花岗岩
水池 / 鹅卵石

内部完成面
入口大厅
天花：石棉吸音板 部分 PB 涂装
墙面：木纹铝格栅
地面：福鼎黑石材地面
休息厅
天花：木纹铝格栅
墙面：木纹铝格栅

清水混凝土
地面：福鼎黑石材地面
观众厅
天花：GRG 吊顶、涂装
墙面：松木防火集成材
地面：高强实木复合木地板
各功能空间：轻钢龙骨石膏板饰面及吊顶

节能
STP 保温板（屋面及外墙）

# Chronology
年表

## 2009

### SEASON 1 / 第一季度

In March, Shanghai Poly Maojia Real Estate Development Co., Ltd. signs a formal contract with Tadao Ando Architect & Associates. Tadao Andao is officially invited as Shanghai Poly Grand Theatre's chief architect. Tadao Ando Architect & Associates commences the first research phase of the theatre's architectural forms.

3月，上海保利茂佳房地产开发有限公司正式与安藤忠雄建筑研究所签订合约，邀请安藤忠雄先生担任上海保利大剧院建筑设计师。安藤忠雄建筑研究所开始建筑形体研究阶段工作

### SEASON 2 / 第二季度

29 April: The proposed construction site is inspected by Tadao Ando and Mayor SUN Jiwei.
30 April: Taodao Ando Architect & Associates submits the mid-term report on its research of architectural forms.
11 June: First presentation of the plans by Tadao Ando to the district government of Jiading. Mayor SUN Jiwei, deputy mayor ZHUANG Mudi, China Poly Group Corporation XUE Ming and LEI Zhongwen, Jiading New City representative LI Jian, planning committee representative GAO Leiping, etc. listen to and discuss the report.
12 June: Tadao Ando presents his architectural plans at the Poly Property Group Co., Ltd.

4月29日，孙继伟区长与安藤忠雄前往基地视察
4月30日，安藤忠雄建筑研究所进行形体研究的中期汇报
6月11日，安藤忠雄先生在嘉定区政府进行第一次方案汇报，孙继伟区长，庄木弟副区长，中国保利集团公司雪明、雷仲文，嘉定新城李俭，规划局高雷平等人听取方案介绍并进行研讨
6月12日，安藤忠雄先生在保利置业做方案汇报

### SEASON 3 / 第三季度

5 August: Second presentation of the plans by Tadao Ando Architect & Associates.
31 August: Third presentation of the plans by Tadao Ando Architect & Associates (the government announces its official approval of the proposal).

8月5日，安藤忠雄建筑研究所第二次方案汇报
8月31日，安藤忠雄建筑研究所第三次方案汇报(政府正式汇报，基本通过方案)

### SEASON 4 / 第四季度

24 November: Shanghai Poly Maojia Real Estate Development Co., Ltd bid on and bought blocks C8-1, C8-2, C11-1, D10-2, D10-14 in Jiading New City's central area. The total area, which totals 151,649 m², includes the commercial and residential districts.
15 December: The first of five research conferences on theatre acoustics is held.

11月24日，上海保利茂佳房地产开发有限公司通过竞买方式获得嘉定新城中心区C8-1、C8-2、C11-1、D10-2、D10-15地块，即商住捆绑地块，共151,649m²土地
12月15日，第一次剧院声学专题研究会议(声学专题会议前后共5次)

## 2010

### SEASON 1

18 January: Press conference on the chosen design for Poly Grand Theatre. The design model was jointly revealed by Jiading District secretary JIN Jianzhong and China Poly Group Corporation deputy general manager XUE Ming.
30 January: The formal design is presented by Tadao Architects & Associates. The breakdown of the area of the Grand Theatre is as follows: 30,000m² for the ground level; 20,000m² for the underground level.
15 March: The project team travels to Japan for the delegation of project tasks and responsibilities. The team inspects the fair-faced concrete process, as well as some of Tadao Ando Architect & Associates' projects.

1月18日，保利大剧院设计方案发布会。嘉定区委书记金建忠、中国保利集团公司副总经理雪明一起为设计方案模型掀起红盖头
1月30日，日本安藤忠雄建筑研究所提出正式设计方案文本，大剧院面积指标分项如下：地上建筑面积约3万m²，地下建筑面积约2万m²
3月15日，项目工作人员赴日进行工作商讨，并考察清水混凝土工艺及安藤忠雄建筑研究所所做项目

### SEASON 2

16 April: The first preliminary in-depth report is presented by Tadao Ando Architect & Associates.
7 May: The Poly Grand Theatre Project's construction management team is established.
26 May: Shanghai Poly Maojia RealEstate Development Co., Ltd signs a contract with the district government of Jiading and the Bureau of Land Management, linking the project grounds with the commercial and residential districts of Jiading.
28 June: The opening ceremony is held at the construction site, marking the official start of construction. The ceremony is attended by China Poly Group Corporation deputy general manager XUE Ming, Jiading district secretary JIN Jianzhong, mayor SUN Jiwei, the director of the Standing Committee, and the chairman of the district committee. The ceremony is hosted by deputy mayor ZHUANG Mudi.

4月16日，安藤忠雄建筑研究所第一次扩初深化汇报
5月7日，"保利大剧院项目建设管理工作小组"成立
5月26日，上海保利茂佳房地产开发有限公司与嘉定区规划和土地管理局签订商住捆绑地块土地出让合同
6月28日，大剧院举行隆重的开工典礼，中国保利集团公司副总经理雪明，上海市委宣传部副部长、嘉定区委书记金建忠，区长孙继伟、区人大主任、区政协主席出席，副区长庄木弟主持开工典礼，大剧院正式进入施工建设阶段

### SEASON 3

16 July: Preliminary assessment of the Grand Theatre is held in Jiading New City Co., Ltd. The meeting is attended by representatives from Shanghai Poly Maojia Co., Ltd and New City Co., Ltd as well as government functionaries and experts from related fields. The assessment is discussed by the participants, and suggestions are made to improve it.
26 September: The second presentation of the preliminary in-depth report is presented by Tadao Ando Architect & Associates.

7月16日，大剧院扩初预评审会在嘉定新城公司召开，上海保利茂佳房地产开发有限公司及新城公司相关负责人，相关政府部门以及各方面专家出席了会议，共同就大剧院扩初设计方案进行讨论，对大剧院扩初设计提出建议
9月26日，安藤忠雄建筑研究所第二次扩初深化汇报

### SEASON 4

11 October: Detailed construction plans of the Grand Theatre are approved by the appropriate authorities.
23 November: The preliminary design review is held in Jiading, and is attended by Shanghai Poly Maojia Real Estate Development Co., Ltd and fourteen government-related organizations.

10月11日，大剧院修详规获得批复
11月23日，大剧院扩初设计评审会在嘉定办证中心召开，上海保利茂佳房地产开发有限公司、设计单位及14家相关政府单位参加

## 2011

### SEASON 1

19 April: A safety training programme is conducted for construction workers.
31 May: Shanghai Poly Grand Theatre receives its construction permits.

4月19日，工人安全技术培训
5月31日，取得施工许可证

### SEASON 2

1 July: Work on the structure's underground level begins.
4 July: The construction team conducts technical studies of Tadao Ando & Associates' projects in Japan.

7月1日，地下结构开始施工
7月4日，现场施工技术团队前往日本对安藤忠雄建筑研究所的项目进行实地现场技术考察

### SEASON 4

3 October: Casting of the basement's roof is complete.
29 October: The construction site is inspected by Tadao Ando and the Japanese team.
1 November: First block of fair-faced concrete is completely poured in.

10月3日，地下室顶板浇筑完毕
10月29日，安藤忠雄先生率日本考察团来剧院工地现场视察
11月1日，第一块清水混凝土浇筑完成

## 2012

5 January: Basement structure passes inspection by the civil quality supervisory committee.
17 March: The construction site is inspected by Tadao Ando.
1月5日，民防地下室结构通过民防质监站验收
3月17日，安藤忠雄先生前往剧院工地现场视察

25 April: Basement structure is approved by Jiading District.
28 June: Safety inspection of the Poly Grand Theatre construction site is attended by over six hundred representatives of various supervisory and construction committees.
4月25日，通过嘉定区地下室结构验收
6月28日，在保利大剧院项目组织嘉定区安全观摩，接待监理及施工单位600余人

5 July: Basement structure passes inspection by Jiading District, and is deemed of high quality.
11 July: Basement structure passes inspection by Shanghai City, and is deemed of high quality.
25 July: Tadao Ando inspects the construction site.
12 August: Onsite operations are directed by SHI Zhiqiang, vice-director of the China Eighth Engineering Bureau.
27 August: Construction begins on the curtain walls.
7月5日，通过嘉定区地下室优质结构验收
7月11日，通过上海市地下室优质结构验收
7月25日，安藤忠雄先生前往剧院工地现场视察
8月12日，中建八局副局长史志强先生现场指导工作
8月27日，幕墙施工人员进场施工

11 October: Site undergoes inspection by the Shanghai municipal government, and is named an "energy-saving" site.
12 November: The theatre's roofing is completed.
10月11日，通过上海市绿色工地验收，并荣获2012年绿色施工（节约型工地）样板工程
11月12日，结构封顶完成

## 2013

11 January: Ground-level structure passes inspection by Jiading District.
16 January: Ground-level structure passes inspection by Jiading District, and is deemed of high quality.
24 January: Ground structure passes inspection by Shanghai City.
3 March: Interior decoration of the theatre is begun.
15 March: Installation of stage machinery is begun.
1月11日，通过嘉定区地上结构验收
1月16日，通过嘉定区地上优质结构验收
1月24日，通过上海市地上优质结构验收
3月3日，装修工程开始施工
3月15日，舞台机械开始安装

In May: The theatre site is named as 2012 Shanghai Civilization and Civilized Model Site.
15 May: Decoration and curtain walls for the construction site are reviewed by architects from Tadao Ando Architect & Associates.
5月，获2012年度上海市文明工地及上海市文明示范工地
5月15日，安藤忠雄建筑研究所建筑师审查施工现场装修及幕墙样板段

18 October: The construction site is inspected by Tadao Ando.
13 December: Work begins on the outdoor plaza's stonework.
10月18日，安藤忠雄先生前往剧院工地现场视察
12月13日，室外广场石材开始铺装

## 2014

14 March: The construction site is inspected by Tadao Ando.
19 March: The construction site is inspected by XU Niansha, CEO of Poly Group.
3月14日，安藤忠雄先生前往剧院工地现场视察
3月19日，保利集团董事长徐念沙到大剧院现场视察

6 August: Fire inspection of the site is conducted.
8 August: The Poly Grand Theatre is visited by XUE Ming, deputy general manager of Poly Group.
11 August: Beginning of the theatre management group's involvement in the project.
26 August: Inspection of the theatre by four agencies.
28 August: Fire safety plan is approved.
5 September: First debugging of the site is held.
9 September: Acoustics test is held.
18 September: Publication of the site's acoustics report.
20 September: Second debugging of the site is held.
22 September: Publication of the supervisory committee's project report.
30 September: Completion of the Shanghai Poly Grand Theatre. The first performance at the theatre is held successfully.
8月6日，消防现场验收
8月8日，保利集团副总经理雪明视察保利大剧院
8月11日，大剧院管理公司进场
8月26日，项目四方验收
8月28日，获取项目消防批复
9月5日，第一场调试演出
9月9日，现场建筑声学测试
9月18日，获取声学检测报告
9月20日，第二场调试演出
9月22日，获取项目验收监督报告
9月30日，上海保利大剧院竣工并成功首演

22 December: "Asian Times", a lecture by Tadao Ando, is delivered in Shanghai.
12月22日，"亚洲的时代"安藤忠雄上海讲演会成功举办

Tadao Ando's sketch
安藤忠雄手绘草图

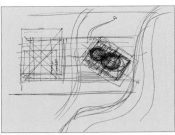
Preliminary sketch
初期草图

Preliminary research of architectural forms
初期形体研究阶段

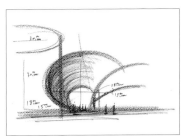
Sketch of spatial relationships
空间关系手绘草图

29 April 2009: Tadao Ando arrives in Shanghai
2009.4.29 安藤忠雄莅临上海

30 April 2009: Mid-term report on the research of architectural forms
2009.4.30 形体研究中期汇报

11 June 2009: First presentation of the plans
2009.6.11 第一次方案汇报

The sketch for the presentation of the plans
方案汇报手绘效果图

1 June 2009: First presentation of the plans
2009.6.11 第一次方案汇报

11 June 2009: Group photo of the first presentation team
2009.6.11 第一次方案汇报相关人员合影

11 June 2009: First presentation of the plans
2009.6.11 第一次方案汇报

11 June 2009: First presentation of the plans
2009.6.11 第一次方案汇报

11 June 2009: Tadao Ando discusses the model with mayor SUN Jiwei
2009.6.11 安藤忠雄与孙继伟区长研讨方案

11 June 2009: Explanation of the model
2009.6.11 方案模型现场演示讲解

11 June 2009: Group photo of Tadao Ando and the CA-GROUP
2009.6.11 安藤忠雄与文筑国际工作人员合影

12 June 2009: Tadao Ando presents the plans to his clients
2009.6.12 向业主方汇报方案

12 June 2009: Tadao Ando makes architectural drawings of the theatre
2009.6.12 安藤忠雄亲自手绘造型示意图

12 June 2009: Tadao Ando presents the plans to his clients
2009.6.12 向业主方汇报方案

12 June 2009: Tadao Ando presents the plans to his clients
2009.6.12 向业主方汇报方案

12 June 2009: architects and clients discuss the models
2009.6.12 设计师与业主方就方案模型进行讨论

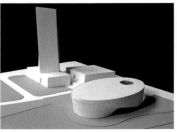
Preliminary model A
初期方案模型 A

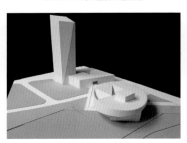
Preliminary model B
初期方案模型 B

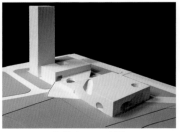
Preliminary model C
初期方案模型 C

18 January 2010: Mayor SUN Jiwei gives a speech during the opening ceremony
2010.1.18 方案揭幕仪式孙继伟区长致词

18 January 2010: Deputy mayor ZHUANG Mudi gives a speech during the opening ceremony
2010.1.18 方案揭幕仪式庄木弟副区长致词

18 January 2010: Architect Kazuya Okano of TAAA gives a speech during the opening ceremony
2010.1.18 方案揭幕仪式安藤事务所冈野一也致词

18 January 2010: Unveiling of the project models
2010.1.18 为方案模型揭幕

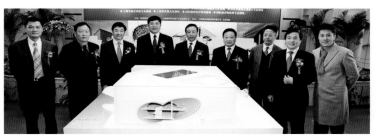
18 January 2010: Group photo of party leaders and guest at the opening ceremony
2010.1.18 方案揭幕仪式上各位领导及嘉宾合影

18 January 2010: Unveiling ceremony
2010.1.18 方案揭幕仪式

18 January 2010: Unveiling ceremony
2010.1.18 方案揭幕仪式

18 January 2010: Project unveiling ceremony
2010.1.18 方案揭幕仪式

22 March 2010: The project design team discusses the plan at TAAA's office in Osaka
2010.3.22 设计团队在大阪安藤事务所讨论细化方案

28 June 2010: Foundation stone ceremony for the Poly Grand Theatre
2010.6.28 保利大剧院奠基典礼

28 June 2010: Mayor SUN Jiwei gives a speech at the foundation stone ceremony
2010.6.28 保利大剧院奠基典礼孙继伟区长发言

26 August 2010: Site inspection of fair-faced concrete
2010.8.26 现场大样混凝土坍落度检查

27 August 2010: Site inspection of structural framework
2010.8.27 现场大样模板检查

19 April 2011: Safety training programme for workers
2011.4.19 工人安全技术培训

26 May 2011: Inspection of fair-faced concrete
2011.5.26 现场大样清水混凝土质量检查

27 May 2011: Inspection of rebars
2011.5.27 基础钢筋捆扎

27 May 2011: Depositing the fair-faced concrete foundation
2011.5.27 基础混凝土浇筑

15 June 2011: Installing the foundational steel beams
2011.6.15 基础筏板钢筋安装

4 July 2011: Construction team visits a construction site in Japan
2011.7.4 施工团队参观日本施工现场

4 July 2011: Construction team visits a construction site in Japan
2011.7.4 施工团队参观日本施工现场

4 July 2011: Construction team visits a construction site in Japan
2011.7.4 施工团队参观日本施工现场

5 July 2011: Construction team discusses plan at TAAA' office in Osaka
2011.7.5 施工团队在大阪安藤事务所讨论施工方案

13 September 2011: Aerial view of the construction site
2011.9.13 俯瞰施工现场

29 October 2011: Tadao Ando inspects the construction site
2011.10.29 安藤忠雄视察工地

29 October 2011: Tadao Ando inspects the construction site
2011.10.29 安藤忠雄视察工地

10 November 2011: Site inspection of of the site framework
2011.11.10 现场模板检查

10 December 2011: Group photo of the project's engineers
2011.12.10 项目工程师合影

17 March 2012: Tadao Ando inspects the construction site
2012.3.17 安藤忠雄视察工地

25 July 2012: Group photo of Tadao Ando and construction workers during site inspection
2012.7.25 安藤忠雄视察工地时与施工人员合影

25 July 2012: Tadao Ando inspects the construction site
2012.7.25 安藤忠雄视察工地

25 July 2012: Tadao Ando inspects the construction site
2012.7.25 安藤忠雄视察工地

25 July 2012: Group photo of Tadao Ando and construction workers during site inspection
2012.7.25 安藤忠雄视察工地时与施工人员合影

25 July 2012: Roof construction
2012.7.25 楼面施工

12 August 2012: Leaders of the China Eighth Engineering Bureau direct work at the construction site
2012.8.12 中建八局领导现场指导工作

29 August 2012: The construction site
2012.8.29 施工现场

11 November 2012: Chairman of the Singapore Construction Office and CIOB Architects Association members conduct project research at the site
2012.11.11 新加坡建设局主席、CIOB 建造师学会成员在施工现场进行项目考察

11 November 2012: Group photo of construction workers and management staff celebrating the completion of the theatre's roof
2012.11.11 结构封顶全体工人和管理人员合影留念

12 November 2012: Photo of Tadao Ando with construction workers
2012.11.12 安藤忠雄与工作人员合影

12 November 2012: Tadao Ando directs work at the construction site
2012.11.12 安藤忠雄现场指导工作

12 November 2012: Completing the roof
2012.11.12 结构封顶

18 April 2013: Aluminum rectangular tube processing site
2013.4.18 圆筒铝方通加工现场

3 May 2013: Construction site interior
2013.5.3 圆筒内部装饰工程现场

15 May 2013: The TAAA inspect samples
2013.5.15 安藤事务所审查装修样板

18 October 2013: Photo of Tadao Ando with construction workers
2013.10.18 安藤忠雄与现场施工人员合影

18 October 2013: Tadao Ando attends a meeting at the construction site office
2013.10.18 施工现场办公室开会

18 October 2013: Tadao Ando's team inspects the construction site
2013.10.18 安藤忠雄一行工地现场视察

18 October 2013: Tadao Ando's team inspects construction site
2013.10.18 安藤忠雄一行工地现场视察

6 March 2014: Photo of Chinese and Japanese staff
2014.3.6 中日双方工作人员合影

14 March 2014: Tadao Ando's team inspects the construction site
2014.3.14 安藤忠雄一行项目视察

14 March 2014: Photo of Tadao Ando with construction workers
2014.3.14 安藤忠雄与工作人员合影

19 March 2014: Heads of Poly Property Group Co., Ltd. inspect the theatre
2014.3.19 保利置业领导视察项目

22 June 2014: Heads of Poly Property Group Co., Ltd. inspect the theatre
2014.6.22 保利置业领导视察项目

8 August 2014: Heads of Poly Property Group Co., Ltd. inspect the theatre
2014.8.8 保利置业领导视察项目

30 September 2014: Chung Myung-whun conducts the first performance at the Shanghai Poly Grand Theatre
2014.9.30 郑明勋上海保利大剧院竣工首演

30 September 2014: The first performance at the Shanghai Poly Grand Theatre was by Chung Myung-whun and the German Philharmonic Orchestra
2014.9.30 上海保利大剧院竣工首演——郑明勋与德国广播爱乐乐团音乐会

30 September 2014: Famous conductor Chung Myung-whun signs autographs at the theatre
2014.9.30 著名指挥家郑明勋现场签名

22 December 2014: Tadao Ando's "Asian Times" lecture in Shanghai
2014.12.22 "亚洲的时代"安藤忠雄上海讲演会

22 December 2014: Tadao Ando chats with mayor SUN Jiwei
2014.12.22 孙继伟区长与安藤忠雄见面会谈

22 December 2014: Group including Tadao Ando and mayor SUN Jiwei visits the Shanghai Poly Grand Theatre
2014.12.22 孙继伟区长与安藤忠雄一行参观上海保利大剧院

22 December 2014: Tadao Ando's "Asian Times" lecture in Shanghai
2014.12.22 "亚洲的时代"安藤忠雄上海讲演会

22 December 2014: Tadao Ando's "Asian Times" lecture in Shanghai
2014.12.22 "亚洲的时代"安藤忠雄上海讲演会

22 December 2014: Tadao Ando's "Asian Times" lecture in Shanghai
2014.12.22 "亚洲的时代"安藤忠雄上海讲演会

22 December 2014: Book-signing
2014.12.22 讲演会签名售书

31 December 2014: The brilliant Shanghai Poly Grand Theatre illuminates Yuanxiang Lake
2014.12.31 褶褶生辉的保利大剧院倒映在远香湖畔

## Credits for Texts

Descriptions of chapter 1: TAAA and CA-GROUP
Descriptions of chapter 2: LI Yi

### Chinese translation from Japanese
CHEN Xudan: pp.24-27
CAO Wenjun: pp.96-103

### English translation from Japanese
Christopher Stephens: pp.24-27, pp.58-61, pp.96-103

### English proofreader
CHOW Lo-Ching

## Credits for Photographs

**Tadao Ando Architect & Associates**
p168, p169(Bottom), pp184-185, pp188-189, pp196-200

**Makoto Yamamori**
Cover, pp8-9, pp10-23, p44, pp60-61, p98, p103, pp104-124, p128, p129(Top), pp130-135, p172, p175(Top), pp176-179, p181, p183, p215, p222

**Yasuhiro Nakayama**
p154, pp158-159, p160, p162, pp164-165, pp171, pp202-207, p213

**Shigeo Ogawa**
pp6-7

**Eiichi Kano**
pp122-123, p129

**CHEN Yan**
p125, p157, pp186-187, p190, p211, p219, p216

**ZHANG Yong**
p182, pp224-225

A+U Publishing Co., Ltd.
Kasumigaseki Building 17F, 3-2-5, Kasumigaseki, Chiyoda-ku,
Tokyo 100-6017, Japan
Tel: +81-3-6205-4384   Fax: +81-3-6205-4387
E-mail: au@japan-architect.co.jp
URL: http://www.japlusu.com

A+U Publishing Pte. Ltd.
Nankin Row, China Square Central
3 Pickering Street, #02-26, Singapore 048660, Singapore
Tel: +65-6557-0537   Fax: +65-6557-0531

© CA-GROUP 2015
Printed in China
Published by A+U Publishing Co., Ltd.
Editorial Department: CA-GROUP (Shanghai)

© CA-GROUP (Shanghai)
No parts of the Magazine, written or pictorial, may be reproduced or published without written permission from the editorial board.

图书在版编目(CIP)数据

安藤忠雄：上海保利大剧院 / 马卫东主编. —上海：同济大学出版社，2015.7
ISBN 978-7-5608-5896-8

Ⅰ.①安… Ⅱ.①马… Ⅲ.①剧院－建筑设计－上海市 Ⅳ.①TU242.2

中国版本图书馆CIP数据核字(2015)第157077号

出版发行：同济大学出版社 www.tongjipress.com.cn
（上海市四平路1239号 邮编：200092 电话：021-65985622）
经　销：全国各地新华书店、建筑书店、网络书店
排版制作：文筑国际
印　刷：上海雅昌艺术印刷有限公司
开　本：889mm×1194 mm 1/16
印　张：14.5
字　数：464千字
版　次：2015年7月第1版　2015年7月第1次印刷
书　号：ISBN 978-7-5608-5896-8
定　价：280.00元

版权所有　侵权必究　印装问题　负责调换